WORKBOOKS IN CHEMISTRY
PHYSICAL CHEMISTRY

Joanne Elliott
University of Reading

Elizabeth Page
University of Reading

Series editor:
Elizabeth Page
University of Reading

OXFORD
UNIVERSITY PRESS

Great Clarendon Street, Oxford, OX2 6DP,
United Kingdom

Oxford University Press is a department of the University of Oxford.
It furthers the University's objective of excellence in research, scholarship,
and education by publishing worldwide. Oxford is a registered trade mark of
Oxford University Press in the UK and in certain other countries

© Joanne Elliott and Elizabeth Page 2017

The moral rights of the authors have been asserted

Impression: 1

All rights reserved. No part of this publication may be reproduced, stored in
a retrieval system, or transmitted, in any form or by any means, without the
prior permission in writing of Oxford University Press, or as expressly permitted
by law, by licence or under terms agreed with the appropriate reprographics
rights organization. Enquiries concerning reproduction outside the scope of the
above should be sent to the Rights Department, Oxford University Press, at the
address above

You must not circulate this work in any other form
and you must impose this same condition on any acquirer

Published in the United States of America by Oxford University Press
198 Madison Avenue, New York, NY 10016, United States of America

British Library Cataloguing in Publication Data

Data available

ISBN 978-0-19-872949-5

Printed in Great Britain by
Bell & Bain Ltd., Glasgow

Links to third party websites are provided by Oxford in good faith and
for information only. Oxford disclaims any responsibility for the materials
contained in any third party website referenced in this work.

Periodic table of the elements

Group	1	2		3	4	5	6	7	8	9	10	11	12	13	14	15	16	17	18
	s-block																		
Period 1	1 H 1.0079																		2 He 4.0026
Period 2	3 Li 6.941	4 Be 9.0122												5 B 10.811	6 C 12.011	7 N 14.007	8 O 15.999	9 F 18.998	10 Ne 20.180
Period 3	11 Na 22.990	12 Mg 24.305												13 Al 26.982	14 Si 28.086	15 P 30.974	16 S 32.065	17 Cl 35.453	18 Ar 39.948
Period 4	19 K 39.098	20 Ca 40.078		21 Sc 44.956	22 Ti 47.867	23 V 50.942	24 Cr 51.996	25 Mn 54.938	26 Fe 55.845	27 Co 58.933	28 Ni 58.693	29 Cu 63.546	30 Zn 65.409	31 Ga 69.723	32 Ge 72.64	33 As 74.922	34 Se 78.96	35 Br 79.904	36 Kr 83.798
Period 5	37 Rb 85.468	38 Sr 87.62		39 Y 88.906	40 Zr 91.224	41 Nb 92.906	42 Mo 95.94	43 Tc (98)	44 Ru 101.07	45 Rh 102.91	46 Pd 106.42	47 Ag 107.87	48 Cd 112.41	49 In 114.82	50 Sn 118.71	51 Sb 121.76	52 Te 127.60	53 I 126.90	54 Xe 131.29
Period 6	55 Cs 132.91	56 Ba 137.33		57 La 138.91	72 Hf 178.49	73 Ta 180.95	74 W 183.84	75 Re 186.21	76 Os 190.23	77 Ir 192.22	78 Pt 195.08	79 Au 196.97	80 Hg 200.59	81 Tl 204.38	82 Pb 207.2	83 Bi 208.98	84 Po (209)	85 At (210)	86 Rn (222)
Period 7	87 Fr (223)	88 Ra (226)		89 Ac (227)	104 Rf (263)	105 Db (262)	106 Sg (266)	107 Bh (272)	108 Hs (277)	109 Mt (276)	110 Ds (281)	111 Rg (280)	112 Cn (277)	113 Nh unknown	114 Fl (289)	115 Mc unknown	116 Lv (298)	117 Ts unknown	118 Og unknown

d-block · p-block

Lanthanides (6)

| 58
Ce
140.12 | 59
Pr
140.91 | 60
Nd
144.24 | 61
Pm
(145) | 62
Sm
150.36 | 63
Eu
151.96 | 64
Gd
157.25 | 65
Tb
158.93 | 66
Dy
162.50 | 67
Ho
164.93 | 68
Er
167.26 | 69
Tm
168.93 | 70
Yb
173.04 | 71
Lu
174.97 |

Actinides (7)

| 90
Th
232.04 | 91
Pa
231.04 | 92
U
238.03 | 93
Np
(237) | 94
Pu
(244) | 95
Am
(243) | 96
Cm
(247) | 97
Bk
(247) | 98
Cf
(251) | 99
Es
(252) | 100
Fm
(257) | 101
Md
(258) | 102
No
(259) | 103
Lr
(262) |

f-block

Key: 8 = Atomic number, O = Symbol, 15.999 = Relative atomic mass

Table of constants and other useful information

Physical constants

Name	Symbol	Value
Avogadro constant	N_A	6.022×10^{23} mol^{-1}
Ideal gas constant	R	8.314 J K^{-1} mol^{-1}
Boltzmann constant	k_B	1.381×10^{-23} J K^{-1}
Planck constant	h	6.626×10^{-34} J s
Faraday constant	F	96 485 C mol^{-1}
Rydberg constant	R_H	1.097×10^7 m^{-1}
Kapustinskii constant	k	1.0790×10^{-4} J m mol^{-1}
		107 900 kJ pm mol^{-1}
Speed of light in vacuum	c	2.998×10^8 m s^{-1}
Elementary charge	e	1.602×10^{-19} C
Electron mass	m_e	9.109×10^{-31} kg
Proton mass	m_p	1.673×10^{-27} kg
Neutron mass	m_n	1.675×10^{-27} kg
Permittivity of free space	ε_0	8.854×10^{-12} J^{-1} C^2 m^{-1}

SI base units

Quantity	Unit name	Symbol
Length	metre	m
Mass	kilogram	kg
Time	second	s
Electric current	ampere	A
Temperature	kelvin	K
Amount of substance	mole	mol
Luminous intensity	candela	cd

Derived units

Quantity	Unit name	Symbol and definition
Area	square metre	m^2
Volume	cubic metre	m^3
Velocity	metre per second	$m\ s^{-1}$
Acceleration	metre per second squared	$m\ s^{-2}$
Density	kilogram per cubic metre	$kg\ m^{-3}$
Concentration	mole per cubic metre	$mol\ m^{-3}$
Energy	joule	$J = kg\ m^2\ s^{-2}$
Force	newton	$N = J\ m^{-1} = kg\ m\ s^{-2}$
Pressure	pascal	$Pa = N\ m^{-2}$
Electric charge	coulomb	$C = A\ s$
Potential difference	volt	$V = J\ C^{-1}$
Power	watt	$W = J\ s^{-1} = kg\ m^2\ s^{-3}$
Frequency	hertz	$Hz\ (s^{-1})$

Multiples of units and prefixes

Multiple	Prefix	Symbol	Multiple	Prefix	symbol
10^{-1}	deci	d	10^1	deca	da
10^{-2}	centi	c	10^2	hecto	h
10^{-3}	milli	m	10^3	kilo	k
10^{-6}	micro	µ	10^6	mega	M
10^{-9}	nano	n	10^9	giga	G
10^{-12}	pico	p	10^{12}	tera	T

Overview of contents

Preface ix

1 Fundamentals

1.1	SI units	1
1.2	Large and small numbers	2
1.3	Converting between units	4
1.4	Moles of atoms and relative atomic masses	5
1.5	Molar mass and determining the amount of substance	7
1.6	Determination of empirical and molecular formula	9
1.7	Gravimetric analysis calculations	10
1.8	Concentrations of solutions	12
1.9	Volumetric analysis	15
1.10	Limiting reagents and percentage yield	18
1.11	The gas laws	19

2 Thermodynamics

2.1	Energy changes	25
2.2	Internal energy, U	27
2.3	Enthalpy changes	31
2.4	Hess's Law	33
2.5	The Born–Haber cycle	37
2.6	Bond enthalpies	40
2.7	Heat capacity and calorimetry	42
2.8	Entropy and the Second Law of Thermodynamics	46

3 Chemical equilibrium

3.1	Equilibrium and the Gibbs energy minimum	58
3.2	The equilibrium constant—approximations for real systems	59
3.3	How K_p, K_c, and K are related	62
3.4	Forward and reverse reactions	63
3.5	Le Chatelier's principle	64
3.6	Standard Gibbs energy change and the position of equilibria	65
3.7	Temperature effect	66
3.8	Calculating equilibrium constants and equilibrium compositions	68
3.9	Solubility	74
3.10	Acids, bases, and water	76
3.11	Ligand substitution reactions	80

4 Phase equilibrium

4.1	Gaseous, liquid, and solid phases	83
4.2	One-component systems—phase behaviour	88
4.3	One-component systems—Gibbs energy, enthalpy, and entropy	90
4.4	One-component systems—Clapeyron equation	94
4.5	One-component systems—Clausius–Clapeyron equation	95
4.6	Two-component mixtures—non-volatile solute plus volatile solvent	98
4.7	Two-component mixtures—ideal binary liquid mixtures	102
4.8	Two-component mixtures—ideal dilute solutions	107
4.9	Two-component mixtures—non-ideal binary liquid mixtures	109
4.10	Two-component mixtures—distillation of binary liquid mixture	111

5 Reaction kinetics

5.1	The rate of a chemical reaction	114
5.2	The order of a chemical reaction	118
5.3	Initial rates method for determining the rate equation	120
5.4	Integrated rate equations	124
5.5	The half-life of chemical reactions	132
5.6	Dependence of rate of reaction upon temperature—the Arrhenius equation	138
5.7	Reaction mechanisms and rate dependence	143

6 Electrochemistry

6.1	Electrostatic interactions	151
6.2	Activity	152
6.3	Ionic strength	154
6.4	Debye–Hückel limiting law	156
6.5	Conductivity	157
6.6	Electrochemical cells	160

Synoptic questions	170
Answers	176
Index	179

Preface

Welcome to the Workbooks in Chemistry

The Workbooks in Chemistry have been designed to offer additional support to help you make the transition from school to university-level chemistry. They will also be useful if you are studying for related degrees, such as biochemistry, food science, or pharmacy.

Introduction to the Workbooks

The Workbooks cover the three traditional areas of chemistry: inorganic, organic, and physical. They are designed to complement your first year chemistry modules and to supplement, but not replace, your course textbook and lecture notes. You may want to use the Workbooks as self-test guides as you carry out a specific topic, or you may find them useful when you have finished a topic as you prepare for end of semester tests and exams. When preparing for tests and exams, students often use practice questions, but model answers are not always available. This is because there is usually more than one correct way to answer a question and your lecturers will want to give you credit for your problem-solving approach and working, as well as having obtained the correct answer. These Workbooks will give you guidance on good practice and a logical approach to problem solving, with plenty of hints and tips on how to avoid typical pitfalls.

Structure of the Workbooks

Each of the three Workbooks is divided into chapters covering the different topics that appear in typical first year chemistry courses. As external examiners and assessors at different UK and international universities, we realize that every chemistry programme is slightly different, so you may find that some topics are covered in more depth than you require, or that there are topics missing from your particular course. If this is the case, we would be interested in hearing your views! However, we are confident that the topics covered are representative, and that most first year students will meet them at some point.

Each chapter is divided into sections, and each section starts with a brief introduction to the theory behind the concepts to put the subsequent problems in context. If you need to, you should refer to your lecture notes and textbooks at this point to fully revise the theory.

Following the outline introduction to each topic, there are a series of **worked examples**, which are typical of the problems you might be asked to solve in workshops or exams. These examples contain fully-worked solutions that are designed to give you the scaffolding upon which to base any future answers, and sometimes provide you with hints about how to approach these types of question and how to avoid common errors.

After the worked examples relating to a topic, you will find further **questions** of a similar type for you to practise. The numerical or 'short' answers to these problems can be found at the end of the book, while fully worked solutions are available on the Online Resource Centre. At the end of each book is a bank of **synoptic questions**, also with worked solutions on the Online Resource Centre. Synoptic questions encourage you to draw on concepts from multiple topics, helping you to use your broader chemical knowledge to solve problems.

You can find the series website at www.oxfordtextbooks.co.uk/orc/chemworkbooks.

How to use the Workbooks

You will probably refer to these Workbooks at different times during your first year course, but we envisage they will be most useful when preparing for examinations after you have done some initial revision.

It is a good idea to use the introductions to the topics to check your understanding and refresh your memory. The next step is to follow through the worked examples, or try them out yourself.

The **comments** will typically relate to the worked solutions and might explain why a unit conversion has been used, for example, or give some background explanation for the maths used in the solution. The comments are designed to help you avoid the typical mistakes students make when approaching each particular type of problem. It is to be hoped that by being aware of these pitfalls you will be able to overcome them.

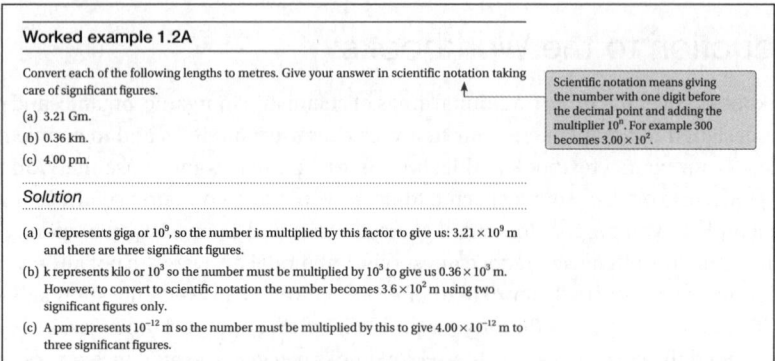

When you are happy you have mastered the worked examples, try the questions. To check your answers go to the back of the book, and to check your working look for the fully worked solutions at www.oxfordtextbooks.co.uk/orc/chemworkbooks.

The synoptic questions can be used as a final revision tool when you are confident with your understanding of the individual topics and want some final practice before the exam or test. Again, you will find answers at the back of the book and full solutions online.

Final comments

We hope you find these workbooks helpful in reinforcing your understanding of key concepts in chemistry and providing tips and techniques that will stay with you for the rest of your chemistry degree course. If you have any feedback on the Workbooks—such as aspects you found particularly helpful or areas you felt were missing—please get in touch with us via the Online Resource Centre. Go to www.oxfordtextbooks.co.uk/orc/chemworkbooks.

1
Fundamentals

1.1 SI units

Units are fundamental to the understanding and application of physical chemistry. The system that has been in use since 1970 is the SI (Système International d'Unités) system of units. There are seven base SI units, of which six are commonly used in chemistry. These are listed in Table 1.1. They are the units of the base quantities of mass, length, time, electrical current, temperature, and amount of substance. The seventh unit is that of luminous intensity which is only rarely relevant in chemistry.

The units for all other physical quantities are derived from these base units. For example, the speed of a moving particle is defined as the distance moved (length) divided by the time taken (time). So the unit of speed, which is defined as $\frac{\text{distance}}{\text{time}}$, is $\frac{\text{m}}{\text{s}}$ or m s^{-1}.

Worked example 1.1A

The SI unit of energy is the joule, J. Determine the derived units for kinetic energy given the expression for kinetic energy of a particle = $\tfrac{1}{2}mv^2$, where m is the mass of the particle and v its velocity.

Solution

As we are given the formula for kinetic energy we can insert the units for the quantities in the equation. Ignoring the value ½, which is dimensionless, we get the following set of units: mv^2 = (kg)(m s^{-1})^2 = kg m^2 s^{-2}.

Worked example 1.1B

Show that the unit for potential energy is equivalent to that of kinetic energy given the expression for the potential energy of a particle is $m \times g \times h$, where m is the mass of the particle, g the gravitational field strength, and h its vertical height.

Table 1.1 Base SI quantities with symbols and units.

Physical quantity	Symbol	Base unit	Unit symbol
Mass	m	kilogram	kg
Length	l	metre	m
Time	t	second	s
Electrical current	I	ampere	A
Temperature	T	kelvin	K
Amount of substance	n	mole	mol

Solution

We are given the expression for potential energy, *PE*, as: $PE = m \times g \times h$.

Mass is a base quantity with the unit of kg. Height is a distance so the unit is m. The symbol *g*, represents gravitational acceleration. So next we need to have the derived units for acceleration. Acceleration is defined as the change in velocity of an object divided by time taken $= \frac{v}{t}$. So the derived units for acceleration are: $\frac{m\,s^{-1}}{s} = m\,s^{-2}$. We are now in a position to combine these units in the expression for potential energy: $PE = m \times g \times h$.

Inserting units for each quantity into the equation gives us:

$$PE = kg \times m\,s^{-2} \times m = kg\,m^2\,s^{-2}$$

Fortunately, this is the same as the derived unit for kinetic energy, and is equivalent to the joule, J.

→ Although these type of manipulations seem a long way from chemistry, it is very useful to be able to carry them out confidently as there are often occasions when you have to convert units into base units in order to simplify an expression. When you need to make these conversions in the problems in this book we will take you through the procedure stepwise.

Question 1.1
Determine the SI unit of density.

Question 1.2
The SI unit for force is the newton, N. Convert this unit into base units given that the definition of force is the mass of an object, *m*, multiplied by its acceleration, *a*.

Question 1.3
There are many different units used for the pressure of a gas. The SI unit of pressure is the pascal and it is defined as the pressure obtained when 1 newton of force is applied over an area of 1 square metre. Determine the base units equivalent to one pascal.

1.2 Large and small numbers

Some values we use in chemistry are extremely large, the Avogadro constant (N_A) for example which is 6.022×10^{23} mol^{-1}, and some are extremely small, for example the Planck constant (*h*) which is 6.626×10^{-34} J s. Scientists therefore tend to use prefixes to represent multipliers which convert the number by the value represented by the prefix. The prefixes kilo, k, and milli, m, are probably the most common and a list of regularly used prefixes and the factors they represent is given in Table 1.2.

It is important to remember to convert units when they have prefixes.

Table 1.2 Most common prefixes used with quantities and units.

Factor	Name	Symbol	Factor	Name	Symbol
10^{12}	tera	T	10^{-2}	centi	c
10^9	giga	G	10^{-3}	milli	m
10^6	mega	M	10^{-6}	micro	μ
10^3	kilo	k	10^{-9}	nano	n
10^{-1}	deci	d	10^{-12}	pico	p

1.2 LARGE AND SMALL NUMBERS

Worked example 1.2A

Convert each of the following lengths to metres. Give your answer in scientific notation taking care of significant figures.

(a) 3.21 Gm.
(b) 0.36 km.
(c) 4.00 pm.

> Scientific notation means giving the number with one digit before the decimal point and adding the multiplier 10^n. For example 300 becomes 3.00×10^2.

Solution

(a) G represents giga or 10^9, so the number is multiplied by this factor to give us: 3.21×10^9 m and there are three significant figures.
(b) k represents kilo or 10^3 so the number must be multiplied by 10^3 to give us 0.36×10^3 m. However, to convert to scientific notation the number becomes 3.6×10^2 m using two significant figures only.
(c) A pm represents 10^{-12} m so the number must be multiplied by this to give 4.00×10^{-12} m to three significant figures.

Worked example 1.2B

Express the following quantities in the units stated:

(a) 1 dm^3 in m^3.
(b) 1 mol dm^{-3} in mol m^{-3}.
(c) 10 m^3 in dm^3.

Solution

When converting between units such as in these questions always think whether your answer is going to be bigger or smaller than the original value.

(a) 1 dm is 0.1 m or 10^{-1} m so 1 dm^3 is $(10^{-1})^3$ $m^3 = 1 \times 10^{-3}$ m^3.
(b) The unit states that we have 1 mol in 1 dm^3 and we are asked to calculate how many mol in 1 m^3. Stop and think whether 1 m^3 is bigger or smaller than 1 dm^3!

It is to be hoped that you decided 1 m^3 is a lot bigger (in fact one thousand times bigger) than 1 dm^3. So if we have just one mol in 1 dm^3 we will have a lot more (in fact 1000 more) in 1 m^3. So a concentration of 1 mol dm^{-3} is the same as 1000 mol m^{-3} or 1×10^3 mol m^{-3}.

(c) This question is asking how many dm^3 are in 10 m^3. We know there are 10 dm in 1 m, and 10^3 dm^3 in 1 m^3. So in 10 m^3 there will be 10×10^3 dm^3. In scientific notation this is 1×10^4 dm^3.

> As 1 dm is smaller than 1 m, 1 dm^3 will be 10^3 times smaller, hence you can check your answer.

> If you find this difficult to envisage think about 1 litre cartons of orange juice in a supermarket. One litre is the same volume as 1 dm^3. So in a space of 1 litre (or 1 dm^3) we have just one carton of orange juice. Imagine now a cubic box with sides of 1 m. This will have a volume of 1 m^3. If you pack 1 litre orange juice cartons into this box you should be able to fit 1000.

❓ Question 1.4

Express the following quantities in the units stated:

(a) 100 dm^3 in m^3.
(b) 10 mol dm^{-3} in mol m^{-3}.
(c) 10 mol dm^{-3} in mol cm^{-3}.
(d) 10^5 mol m^{-3} in mol dm^{-3}.

Question 1.5

Convert each length to metres; record your answer in scientific notation and take care of significant figures.

(a) 2002 mm.

(b) 35 dm.

(c) 295.0 µm.

1.3 Converting between units

One of the commonest sources of errors in chemical calculations is failing to convert values into similar units in an expression. For example, the SI unit of mass is the kilogram. But we typically weigh substances in gram quantities because this is a more convenient amount. However, as we have seen, many SI units—for example, energy, force, pressure, etc.—involve the kilogram, and so it is important to remember to convert masses into kilograms where this is relevant.

Worked example 1.3A

Calculate the pressure in pascal exerted by 100 g mercury in a manometer of cross section 1 cm^2.

Solution

The pascal, the unit of pressure, is defined as the force exerted by an object divided by the area it acts upon.

$$\text{pressure} = \frac{\text{force}}{\text{area}}.$$

The mercury in this barometer will have a force due to its mass and the gravitational acceleration, g which equals 9.8 m s^{-2}. So the equation for pressure becomes: $\text{pressure} = \frac{\text{mass} \times g}{\text{area}}$.

Now we have the expression for pressure with the appropriate quantities included we have to convert all our quantities to SI base units to ensure the final value is in pascal. The mass of the mercury is 100 g which is equivalent to 0.1 kg. The cross-sectional area of the manometer is given as 1 cm^2.

To convert from g to kg divide by 1000 or multiply by 10^{-3}.

1 cm = 0.01 m or 10^{-2} m.

So 1 cm^2 = (0.01)^2 m^2 or (10^{-2})^2 m^2 = 10^{-4} m^2.

Now we are in a position to put the values into the expression for pressure:

$$\text{Pressure} = \frac{0.1 \text{kg} \times 9.8 \text{ m s}^{-2}}{10^{-4} \text{ m}^2} = 9.8 \times 10^3 \text{ kg m}^{-1} \text{ s}^{-2} = 9.8 \times 10^3 \text{ Pa}$$

m/m^2 = m^{-1} and the resulting unit of kg m^{-1} s^{-2} = Pa.

Worked example 1.3B

The wavelength of light, λ, is related to its frequency, υ, and speed, c, by the equation $\lambda = \frac{c}{\upsilon}$. If the wavelength of a monochromatic ray of red light is 650 nm calculate the frequency in SI units given the speed of light is 3.00×10^8 m s^{-1}.

Solution

The SI unit of frequency is s^{-1} (or Hz). So to obtain the frequency in s^{-1} we must convert the wavelength to SI units by multiplying by 10^{-9} as the wavelength is given in nanometres. Thus λ becomes 650×10^{-9} m.

We must also rearrange the equation to make the frequency, ν, the subject:

➔ $1\ s^{-1} = 1\ Hz$

$$\lambda = \frac{c}{\nu} \quad \text{so} \quad \nu = \frac{c}{\lambda} = \frac{3.00 \times 10^8\ m\ s^{-1}}{650 \times 10^{-9}\ m} = 4.62 \times 10^{14}\ s^{-1}$$

So the answer is $4.62 \times 10^{14}\ s^{-1}$.

The unit of m cancels on both top and bottom of the equation leaving a unit of s^{-1}.

This answer correctly gives the value to three significant figures and is in scientific notation.

Question 1.6

A certain line in the electromagnetic spectrum has the frequency 7.25 THz. Calculate the energy in J of one photon having this frequency given that the relationship between energy and frequency is given by the equation $E = h\nu$, where E is the energy, ν the frequency, and h is the Planck constant $= 6.626 \times 10^{-34}$ J s. Calculate the energy in kJ of one mole of such photons.

The ideal gas law relates the pressure, p, volume, V, temperature, T, and amount of a gas, n, and includes the gas constant, $R = 8.3145$ J K^{-1} mol^{-1}. The equation can be represented by: $pV = nRT$. The various units used for pressure, volume, and temperature can lead to some mathematical acrobatics!

Question 1.7

Calculate the volume in dm^3 occupied by 10 mol of oxygen at 1.0 atm and 25 °C using the ideal gas law, $pV = nRT$ and a value of 8.3145 J K^{-1} mol^{-1} for the gas constant, R.

1.4 Moles of atoms and relative atomic masses

The **mole** is defined as the number of atoms in exactly 12 g of carbon-12 (^{12}C). This number is defined as the **Avogadros constant**, N_A, and has a value of 6.022×10^{23} mol^{-1}. Although the Avogadro constant was derived in terms of the number of atoms in one mole of a substance, the constant can relate to any number of items, for example molecules, ions, photons, electrons, etc.

An atom of the isotope ^{12}C has a **relative atomic mass**, A_r, of exactly 12. The relative atomic mass in grams of any element contains 6.022×10^{23} atoms of that element. So, for example, 1.0 g of ^{1}H contains 6.022×10^{23} atoms of ^{1}H, and 1.0 g of ^{1}H$_2$ contains half a mole of ^{1}H$_2$ or 3.011×10^{23} molecules of ^{1}H$_2$ gas.

We generally obtain values for relative atomic masses from the periodic table. However, if you check the periodic table (see page iii) you will see that hydrogen has an A_r of 1.0079 not 1.000 and carbon has A_r of 12.011 not 12.000. In fact, most relative atomic masses of elements are not whole numbers. This is because a naturally occurring sample of an element contains several different isotopes of the element, each with its own relative abundance. So in a sample of hydrogen gas we have ^{1}H, ^{2}H (deuterium), and a very tiny amount of ^{3}H (tritium). This makes the average mass of an atom of hydrogen slightly higher than 1.000 g and the same is true for carbon which is composed of ^{12}C, ^{13}C, and ^{14}C—thus making the average relative atomic mass of C 12.011 not 12.000.

Worked example 1.4A

How many atoms of (i) hydrogen and (ii) oxygen are there in 18.0 g of water?

Solution

The first step is to find the molar mass of water from the formula, H_2O.

$$2 \times H\,(1.00\text{ g}) + 1 \times O\,(16.0\text{ g}) = 18.0\text{ g}$$

> As we only know the mass of water to 3 significant figures it is not appropriate to use 5 figure accuracy for the atomic masses.

We therefore have 1 mole of water molecules. 1 mole of anything contains Avogadro's number, i.e. 6.022×10^{23} units.

Number of H_2O molecules in one mole = 6.022×10^{23}.
In one H_2O molecule there are 2 H atoms and 1 O atom so:
Number of H atoms in 18.0 g water = $6.022 \times 10^{23} \times 2 = 12.04 \times 10^{23}$ or 1.204×10^{24}.
Number of O atoms in 18.0 g water = 6.02×10^{23} (to 3 significant figures).

> ➔ Molar mass, M, is the mass per mole (g mol^{-1}) of a substance. This can be atoms, molecules, or formula units (see p. 7).

Worked example 1.4B

How many (i) sodium ions and (ii) phosphate ions are there in 10 moles of sodium phosphate, Na_3PO_4?

Solution

1 mole Na_3PO_4 contains 3 moles Na^+ ions and 1 mole PO_4^{3-} ions. Therefore 10 moles will contain $3 \times 10 = 30$ moles Na^+ ions and 10 moles PO_4^{3-} ions.

As 1 mole of a substance contains N_A, 6.022×10^{23}, units:
30 moles Na^+ ions = $30 \times 6.022 \times 10^{23}$ ions = 180.66×10^{23} ions = 1.8×10^{25} ions.
10 moles PO_4^{3-} ions = $10 \times 6.022 \times 10^{23} = 6.0 \times 10^{24}$ ions.

Worked example 1.4C

A naturally occurring sample of iron metal is made up of isotopes of iron with the following relative abundances:

$^{54}Fe = 5.86\%$, mass of atom = 53.9 u
$^{56}Fe = 91.8\%$, mass of atom = 55.9 u
$^{57}Fe = 2.12\%$, mass of atom = 56.9 u
$^{58}Fe = 0.22\%$, mass of atom = 57.9 u

Calculate the relative atomic mass of iron.

> ➔ The atomic mass unit, u, is defined as one twelfth of the mass of a carbon-12 atom and is used to express masses of atomic particles, u = 1.661×10^{-27} kg. The symbols Da (dalton) and amu (atomic mass unit) are sometimes used.

Solution

To solve a problem such as this consider a sample of iron containing 100 atoms. The distribution of the iron isotopes in these 100 atoms will be according to their relative abundances. We can therefore calculate the mass of 100 such atoms and then take the average.

Mass of 100 iron atoms = $(53.9 \times 5.86 + 55.9 \times 91.8 + 56.9 \times 2.12 + 57.9 \times 0.22)$ u = $(312.6 + 5131.6 + 120.6 + 12.75)$ u = 5577.5 u

Therefore mass of 1 atom = $5577.5/100$ u = 55.8 u.
To three significant figures this gives us a relative atomic mass for iron of 55.8.

> **Question 1.8**
>
> How many (i) nitrogen atoms and (ii) nitrogen molecules are there in 14.0 g nitrogen gas?

Question 1.9

Determine the total number of ions formed in solution when 1 mole of aluminium sulfide dissociates completely.

 Question 1.10

Magnesium has three naturally occurring isotopes with the following relative abundances: $^{24}Mg = 78.6\%$, $^{25}Mg = 10.1\%$, and $^{26}Mg = 11.3\%$. Calculate the relative atomic mass of magnesium assuming the relative isotopic masses of Mg are 24.0, 25.0, and 26.0 respectively.

1.5 Molar mass and determining the amount of substance

The formula of a substance tells us the number and types of atoms in the compound, whether it be a molecule or an ionic species—as we have already encountered in some of the earlier examples. The mass of one mole of the substance is calculated by adding together the relative atomic masses of all the atoms shown in the formula. This is called the **relative formula mass** or **relative molecular mass** of the substance, M_r. So the mass of one mole of a substance is equal to the relative formula mass in grams. We call this the **molar mass** of the substance, and represent it by the symbol M.

We can use the molar mass to calculate the amount in moles of a substance present if we know its mass and formula, and can use it to calculate the **mass** of substance if we know the amount in moles present. The equation that links the quantities is: $n = \dfrac{m}{M}$ where n is the amount in moles of the substance, m is the mass of substance, and M is the molar mass. The molar mass has the units g mol^{-1}.

Worked example 1.5A

Calculate the mass of two moles of ethanol, C_2H_5OH.

Solution

A problem of this type can be solved in two different ways—although they basically involve the same procedure.

The question gives us the formula for ethanol, C_2H_5OH, so we can use a knowledge of relative atomic masses, or data given in the periodic table, to work out the relative mass of one formula unit of ethanol and therefore its molar mass:

$M_r(C_2H_5OH) = (2 \times 12.01) + (5 \times 1.008) + (1 \times 16.00) + (1 \times 1.008) = 46.07$.

So the molar mass of C_2H_5OH is 46.07 g mol^{-1}. The mass of **two** moles of C_2H_5OH is therefore $2 \text{ mol} \times 46.07 \text{ g mol}^{-1} = 92.14$ g.

Alternatively you can rearrange the formula $n = \dfrac{m}{M}$ to make the mass, m, the subject of the equation: $m = n \times M$. From here the procedure is exactly as above: we find the molar mass, M, of ethanol and then multiply by the number of moles:

$m = 2 \text{ mol} \times 46.07 \text{ g mol}^{-1} = 92.14$ g.

Worked example 1.5B

Calculate the amount in moles in 1 kg of urea, $OC(NH_2)_2$.

→ The relative atomic masses are given here to four significant figures. The values you use for atomic masses will depend upon the situation—i.e. whether the problem is one which requires a good level of accuracy such as in quantitative work or whether you are looking for an approximate amount to weigh out in a synthesis in which case two figure accuracy will suffice. If you are not certain from the question then it is always safer to use four significant figures.

→ When posed in this manner such questions are relatively straightforward to answer. However, these types of calculations, which involve masses and moles, are generally part of more complex problems so it is important to be able to master the basics before proceeding to more involved calculations

Notice how the units of mol and mol^{-1} cancel, leaving us with the units of mass as g.

Solution

In this problem we are asked to calculate the amount in moles and so the equation $n = \frac{m}{M}$ is the one to use.

We are given the formula of urea and so must work out the molar mass first.

Molar mass of $OC(NH_2)_2 = ((1 \times 16.00) + (1 \times 12.01) + (2 \times 14.01) + (4 \times 1.01))$ g $= 60.07$ g mol^{-1}.

Substituting the values in the equation:

> Don't forget to convert kg to g by multiplying by 1000.

$$n = \frac{1000 \text{ g}}{60.07 \text{ g mol}^{-1}} = 16.65 \text{ mol}$$

Worked example 1.5C

A 1 litre solution of an alkali metal carbonate was determined to contain 69.1 g of the metal carbonate, which was found by titration to be equivalent to half a mole of the salt. What is the nature of the metal in this salt?

Solution

The question implies that 0.5 mole of the metal carbonate has a mass of 69.1 g. We can therefore find the molar mass of the metal carbonate by rearranging the equation: $n = \frac{m}{M}$ to give $M = \frac{m}{n}$.

Substituting in the values given: $M = \frac{69.1 \text{ g}}{0.5 \text{ mol}}$. So $M = 138.2$ g mol^{-1}.

The question states that this is an alkali metal carbonate so the formula must be M_2CO_3 (where M is the Group 1 metal). The mass of one mole of carbonate ion (CO_3^{2-}) can be calculated to be 60.01 g.

Subtracting the mass of the carbonate ion from the mass of the salt gives us the mass of the metal ions in the salt:

Mass of $(M^+)_2 = (138.2 - 60.01)$ g mol^{-1} = 78.19 g mol^{-1}

Therefore the mass of M^+ ions = 78.19 g mol^{-1}/2 = 39.1 g mol^{-1}. By checking the atomic masses of the Group 1 metals we can see that the metal must be potassium with an atomic mass of 39.1 g mol^{-1}.

> **Question 1.11**
>
> Calculate the mass of 0.50 moles of glucose, $C_6H_{12}O_6$.

> **Question 1.12**
>
> A 1000 ml solution of saline drip is required containing 9.00 g sodium chloride. How many moles of sodium chloride does this contain?

> **Question 1.13**
>
> Treatment of a solution containing halide ions with 0.01 mol silver nitrate resulted in the complete precipitation of the silver halide which had a mass of 1.88 g. Determine the nature of the halide ion in the precipitate.

1.6 Determination of empirical and molecular formula

The **empirical formula** of a compound gives us the **relative numbers** of atoms of each element present in the compound. As such, it gives us information about the ratio of atoms in the compound but **not** the actual number as the empirical formula is not always the same as the molecular formula.

By contrast, the **molecular formula** tells us the actual number of atoms of each element in the compound, from which we can derive the molar mass. For example, glucose has the molecular formula $C_6H_{12}O_6$ which indicates that one molecule of glucose contains 6 C atoms, 12 H atoms, and 6 O atoms. However, the empirical formula of glucose is CH_2O; this is the formula we would obtain from analysis of the carbon and hydrogen content in the substance. With this information, mass spectrometry would generally be used to determine the molecular formula.

Worked example 1.6A

The molecule caffeine was found by CHN analysis to have the following percentage composition by mass: 49.5% C, 5.2% H, 28.9% N. Mass spectrometry showed it to have a molar mass of 194.2 g mol^{-1} for the parent ion. Determine the empirical and molecular formula of caffeine.

Solution

The first point to notice about a problem such as this is that the percentages of C, H, and N do not add up to 100%. In this situation we have to assume that the remainder of the compound is made up of oxygen—as this is not easy to determine analytically. So the percentage of oxygen must be determined by summing the C, H, and N percentages and subtracting from 100%.

Percentage oxygen = (100 − (49.5 + 5.2 + 28.9)) = 16.4%

Now we have the percentages of each element we can find the molar ratios by dividing by the atomic masses. This assumes we have 100 g of the compound and can convert each percentage amount to grams.

$$n = \frac{m}{M}$$

Amount in moles of C in 100 g, $n(C) = \dfrac{49.5 \text{ g}}{12.01 \text{ g mol}^{-1}} = 4.122 \text{ mol}$

Amount in moles of H in 100 g, $n(H) = \dfrac{5.2 \text{ g}}{1.01 \text{ g mol}^{-1}} = 5.149 \text{ mol}$

Amount in moles of N in 100 g, $n(N) = \dfrac{28.9 \text{ g}}{14.01 \text{ g mol}^{-1}} = 2.063 \text{ mol}$

Amount in moles of O in 100 g, $n(O) = \dfrac{16.4 \text{ g}}{16.00 \text{ g mol}^{-1}} = 1.025 \text{ mol}$

To obtain the simplest ratio of the elements in the compound we now divide through by the smallest number of moles. In this case the smallest number is for oxygen which is 1.025 mol in 100 g.

Ratio of C to O = $\dfrac{4.122}{1.025} = 4.021 \approx 4$

Ratio of H to O = $\dfrac{5.149}{1.025} = 5.023 \approx 5$

> We are only looking for ratios of numbers here so the values can be rounded to the nearest whole number.

$$\text{Ratio of N to O} = \frac{2.063}{1.025} = 2.013 \approx 2$$

$$\text{Ratio of O to O} = \frac{1.025}{1.025} = 1.000 = 1$$

So the ratio of the elements in the compound—and hence the empirical formula—is $C_4H_5N_2O$.

Notice that it was not actually necessary in this case to divide through by the smallest number of atoms as the number of moles of oxygen in 100 g was very close to one mole and so could be approximated to one.

The mass of 1 mole of the empirical formula unit is given by summing the atomic masses for each of the elements multiplied by the number of each element:

$$M = (4 \times 12.01) + (5 \times 1.01) + (2 \times 14.01) + (1 \times 16.00) = 97.11 \text{ g mol}^{-1}$$

As we are told the molecular mass of caffeine is 194.2 g mol^{-1} we obtain the number of empirical units of caffeine in the molecular formula by dividing the molar mass by the mass of one empirical unit:

$$\text{Number of empirical units in 194.2 g mol}^{-1} = \frac{194.2 \text{ g mol}^{-1}}{97.11 \text{ g mol}^{-1}} = 2.000 \approx 2$$

So the formula of caffeine is $C_8H_{10}N_4O_2$.

Question 1.14

The compound nicotine has the following elemental analysis: C = 74.02%, H = 8.71%, N = 17.27%. The molecular mass of nicotine is 162.3 g mol^{-1}. Determine the empirical formula and molecular formula of nicotine.

Question 1.15

A green compound is known to contain only chromium and oxygen. It is found to contain 68.42% chromium.

(a) Determine the empirical formula of the compound.

(b) A mass spectrometer shows the molar mass to be 152 g mol^{-1}. Calculate the molecular formula.

1.7 Gravimetric analysis calculations

Gravimetric analysis involves determining the amount of substance present by taking measurements of mass. This type of analysis typically involves precipitating an insoluble compound from solution before filtering, drying, and weighing it. If the nature of the precipitate is known, its M_r can be used along with the mass of precipitate to find the amount in moles. Knowing the amount in moles of precipitate allows us to determine the amount in moles of one of the component atoms or ions in the original substance.

Worked example 1.7A

A sample of a metal chloride was prepared and the chloride content analysed by gravimetric analysis. 6.137 g of the metal chloride was dissolved in water and the solution made up to 250 cm³ in a volumetric flask. 25 cm³ portions of the chloride solution were taken and excess silver nitrate

1.7 GRAVIMETRIC ANALYSIS CALCULATIONS

solution added. The precipitated silver chloride was filtered and dried and was found to weigh 1.669 g. Calculate the percentage chlorine in the original salt.

Solution

The first step in such a calculation is to determine the nature of the precipitate from the stoichiometric equation for the reaction. In this example we have a 1:1 reaction between Ag^+ ions and Cl^- ions to give solid silver chloride:

$$Ag^+ (aq) + Cl^- (aq) \rightarrow AgCl (s)$$

We are told the mass of silver chloride precipitated and so can find the amount in moles.

$$n(AgCl) = \frac{m}{M} = \frac{1.669 \text{ g}}{143.4 \text{ g mol}^{-1}} = 0.01164 \text{ mol}$$

Therefore the amount in moles of Cl^- in 25 cm³ of the original solution is 0.01164. The amount in of moles of Cl^- in the sample is 0.01164×10 as the sample was dissolved in 250 cm³.

> If there are 0.01164 mol Cl^- in 25 cm³ there will be 0.01164×10 mol Cl^- in 250 cm³

So the 6.137 g sample contains 0.1164 mol Cl^- ion.

To find the mass of chloride in the salt we convert the amount in moles to mass of Cl by multiplying by the atomic mass of Cl:

Mass of Cl in sample = $0.1164 \text{ mol} \times 35.45 \text{ g mol}^{-1} = 4.126$ g

The percentage of chlorine in the sample is $\frac{4.126 \text{ g}}{6.137 \text{ g}} \times 100\% = 67.2\%$.

Worked example 1.7B

Nickel can be determined gravimetrically by precipitation with dimethylglyoxime ($DMGH_2$) according to the equation:

$$Ni^{2+} (aq) + 2 \text{ DMGH}_2 (aq) \rightarrow Ni(DMGH)_2 (s) + 2 H^+ (aq)$$

1.2145 g of a nickel(II) salt was weighed accurately and dissolved in 250 cm³ water in a volumetric flask. 25 cm³ portions of the Ni^{2+} solution were taken and excess $DMGH_2$ solution added until no further precipitation occurred. The red precipitate was filtered, washed, and dried. One such determination yielded 0.2256 g precipitate. Determine the percentage of nickel in the original Ni(II) salt.

$DMGH^- = C_4H_7O_2N_2$, $M_r = 115.13$

Solution

From the stoichiometric equation given in the question we know that the composition of the solid precipitate formed is $Ni(DMGH)_2$ or $Ni(C_4H_7O_2N_2)_2$.

To solve this problem we first find the amount in moles of $Ni(DMGH)_2$ precipitated (step 1). This will give us the amount in moles of Ni^{2+} in the 25 cm³ portion. We can then get the total amount in moles of Ni^{2+} in the 250 cm³ portion (step 2) and then the mass of Ni (step 3). As we are given the mass of the original salt used we can find the proportion of the salt that is nickel (step 4).

Step 1: We can obtain the number of moles of $Ni(DMGH)_2$ from the mass divided by the molar mass. We must first determine the molar mass of the complex. The molar mass of $DMGH^-$ is given as 115.13 g mol⁻¹.

The molar mass of $Ni(DMGH)_2 = 58.69 + 2 \times 115.13 = 288.95$ g mol⁻¹

> Use the equation $n = m/M$

$n(Ni(DMGH)_2)$ (Number of moles of $Ni(DMGH)_2$) $= \frac{0.2256 \text{ g}}{288.95 \text{ g mol}^{-1}} = 7.808 \times 10^{-4}$.

Step 2: $n(\text{Ni}^{2+})$ in 25 cm^3 = 7.808 × 10^{-4}. Therefore $n(\text{Ni}^{2+})$ in 250 cm^3 = 10 × (7.808 × 10^{-4}) = 7.808 × 10^{-3}.

Step 3: Mass of Ni^{2+} in 250 cm^3 = $n \times M$ = (7.808 × 10^{-3} mol) × 58.69 g mol^{-1} = 0.4582 g.

Step 4: To calculate the percentage of nickel in the salt we then divide the mass of nickel by the mass of the salt and multiply by 100:

$$\frac{0.4582\,\text{g}}{1.2145\,\text{g}} \times 100\% = 37.73\% = 37.7\%$$

> **Question 1.16**
>
> A 2.100 g sample of an aluminium salt was dissolved in 250 cm^3 distilled water in a volumetric flask. 25.0 cm^3 portions were taken, ethanoic acid added until the pH was around 4.5, and excess 8-hydroxyquinoline added. The resulting yellow precipitate was filtered, washed, and dried and found to weigh 0.5672 g. The yellow precipitate has the formula Al(quin)$_3$. Calculate the percentage aluminium in the original salt.
>
> 8-hydroxyquinoline (quin) = C$_9$H$_7$NO; M_r = 145.16, quin$^-$ = C$_9$H$_7$NO$^-$; M_r = 144.15

> **Question 1.17**
>
> A certain barium halide exists as the hydrated salt BaX$_2$·2H$_2$O, where X is the halogen. The barium content of the salt can be determined by gravimetric methods. A sample of the halide (0.2650 g) was dissolved in water (200 cm^3) and excess sulfuric acid added. The mixture was then heated and held at boiling for 45 minutes. The precipitate (barium sulfate) was filtered off, washed, and dried. Mass of precipitate obtained = 0.2533 g. Determine the identity of X.

1.8 Concentrations of solutions

> 1 dm^3 = 1000 cm^3 or 1 l.
> Try to work in dm^3 so the volume units are consistent and cancel

The unit most commonly used by chemists to define the concentration of a solution is mol dm^{-3}. This is termed the **molarity** of the solution. This unit is generally the most useful one for chemists because it indicates the amount in moles of a substance in a certain volume of solution (normally water).

A simple equation that allows us to calculate molarity is given by $c = \dfrac{n}{V}$ where:

> The unit M is sometimes used in place of mol dm^{-3}

- c is the concentration (in mol dm^{-3}).
- n is the amount in moles.
- V is the volume (expressed in dm^3).

This equation directly results from the expression of concentration as being the amount in moles of substance dissolved in a certain given volume. Very often we must first calculate the amount in moles from the mass of the substance (or solute—the substance being dissolved).

The equation $c = \dfrac{n}{V}$ allows us to calculate concentration but also the amount in moles of a substance in a certain volume of solution. This is particularly important in titration calculations.

The examples below are calculations you may have to make when preparing a standard solution of a solute in a volumetric (or graduated) flask when preparing for a titration.

Worked example 1.8 A

Potassium hydrogen phthalate, KHP (HOOCC$_6$H$_4$COOK, M_r = 204.22), is used as a primary standard in the determination of the concentration of sodium hydroxide solutions. Calculate the mass of KHP that must be weighed to make 250 cm^3 0.1000 M solution.

In practice you would not make a solution with an exact concentration. Generally it is sufficient to have an approximate concentration and know the value accurately.

Solution

This type of problem must be attacked in two stages. The first stage is to use the concentration of the solution required and the volume of the flask to determine the amount in moles of KHP that must be used.

First, rearrange the equation to make the number of moles the subject: $c = \dfrac{n}{V}$ so $n = c \times V$. Substituting in values for c and V:

$$n = 0.1000 \text{ mol dm}^{-3} \times 250 \times 10^{-3} \text{ dm}^3 = 0.0250 \text{ mol}$$

Don't forget to convert the volume in cm^3 to dm^3 by dividing by 1000 or multiplying by 10^{-3} so the units will cancel.

Having determined the amount in moles of KHP required we now need to calculate the mass using $n = \dfrac{m}{M}$ and rearrange the equation to make m the subject.

$$m = n \times M = 0.0250 \text{ mol} \times 204.22 \text{ g mol}^{-1} = 5.11 \text{ g}$$

Worked example 1.8B

In the above procedure to prepare a standard solution of potassium hydrogen phthalate, (HOOCC$_6$H$_4$COOK, M_r = 204.22), the chemist weighed 5.0095 g solid. Calculate the actual molarity of the resulting solution that was prepared.

➔ In practice this is the way the standard solution would be prepared. First, a rough calculation is carried out to find the mass of KHP required and then a sample with this approximate mass would be weighed accurately and dissolved in a known volume of water.

Solution

The procedure here is the reverse of the procedure in Worked example 1.8A. The actual amount in moles weighed is first calculated and then the concentration of the solution when this is dissolved up to 250 cm^3 in the volumetric flask is determined.

Calculation of the amount in moles of KHP:

$$n = \dfrac{m}{M} \text{ so, } n = \dfrac{5.0095 \text{ g}}{204.22 \text{ g mol}^{-1}} = 0.02453 \text{ mol}$$

The amount in moles can now be used to work out the concentration of the standard solution in mol dm^{-3} or M.

$$c = \dfrac{n}{V} \text{ so, } c = \dfrac{0.02453 \text{ mol}}{250 \times 10^{-3} \text{ dm}^3} = 9.81 \times 10^{-2} \text{ mol dm}^{-3} = 0.0981 \text{ mol dm}^{-3}.$$

 Question 1.18

Benzoic acid (C$_6$H$_5$COOH, M_r = 122.12) is a primary standard used for the standardization of ethanolic sodium and potassium hydroxide. Calculate the mass of benzoic acid that must be weighed to produce 100 ml of a 0.010 M standard solution.

Question 1.19

Calculate the molarity of the benzoic acid solution formed if a mass of 0.1311 g solid was weighed and dissolved in a 100 ml volumetric flask.

Question 1.20

An unknown solution of ammonium chloride was standardized using sodium hydroxide solution of concentration 0.0998 M. 50.0 cm^3 of this NaOH solution was added to 25.0 cm^3 of the unknown ammonium chloride and the unreacted sodium hydroxide was titrated with standard HCl. Titration showed that there were 2.025×10^{-3} mol NaOH remaining after complete neutralization of the ammonium chloride solution. Determine the original concentration of the ammonium chloride solution.

Diluting solutions

Very dilute solutions are usually made by diluting more concentrated, **stock**, solutions. This is because the amount in grams of the solute required as the solution becomes weaker becomes very small and the inaccuracies associated with weighing such small masses get bigger. So we start by preparing a concentrated solution by weighing a reasonable mass of the solid and then dilute the solution down by using accurately measured volumes. If a certain volume of a more concentrated solution is diluted to a larger volume, the amount in of moles of solute does not change, but the concentration of the solution is reduced.

Worked example 1.8C

500 cm^3 of a 0.00500 M solution of potassium manganate (VII), KMnO$_4$, is required by dilution of a stock solution of 0.100 M KMnO$_4$. What volume of KMnO$_4$ solution must be taken?

Solution

The key to this type of problem is to remember that the amount in moles of solute (in this case KMnO$_4$) in the volume required must be the same as the amount in moles in the final diluted solution. We can therefore apply the equation: $n_i = n_f$ where i and f denote the initial and final solutions respectively and n is the amount in moles in each.

This relationship can be written as: $c_i \times V_i = c_f \times V_f$.

The volume we need to determine is represented by V_i and so we can insert the known values into the equation and make V_i the subject of the equation. Don't forget to convert the volume to dm^3 in order to work in moles.

It may help to define $c_f = 0.00500$ M, $V_f = 500$ cm^3, and $c_i = 0.100$ M

$$V_i = \frac{c_f \times V_f}{c_i} = \frac{0.00500 \text{ mol dm}^{-3} \times 500 \times 10^{-3} \text{ dm}^3}{0.100 \text{ mol dm}^{-3}} = \frac{2.50 \times 10^{-3} \text{ mol}}{0.100 \text{ mol dm}^{-3}}$$
$$= 2.50 \times 10^{-2} \text{ dm}^3 = 25.0 \times 10^{-3} \text{ dm}^3 = 25.0 \text{ cm}^3$$

Worked example 1.8D

20.0 cm^3 of a standardized solution of 0.5051 M HCl was diluted to 500 cm^3 in a volumetric flask. What is the new concentration of the HCl solution?

Solution

We can use the relationship: $c_i \times V_i = c_f \times V_f$ to solve this problem by first rearranging the equation to make the new concentration the unknown quantity:

$$c_f = \frac{c_i \times V_i}{V_f} = \frac{0.5051 \text{ mol dm}^{-3} \times 20.0 \times 10^{-3} \text{ dm}^3}{500 \times 10^{-3} \text{ dm}^3} = \frac{10.10 \times 10^{-3} \text{ mol}}{500 \times 10^{-3} \text{ dm}^3}$$

$$= 0.0202 \text{ mol dm}^{-3}$$

> This type of problem can also be solved by simple dilutions as in this case we are taking a 20.0 cm³ portion and diluting to 500 cm³. The new concentration will therefore be 20.0/500 times weaker than the original, i.e. a factor of 0.04 more dilute. If we multiply the original concentration by 0.04 we get the new concentration = 0.04×0.5051 mol dm⁻³ = 0.0202 mol dm⁻³.

 Question 1.21

A stock solution of $Na_2S_2O_3$ is 0.500 M. A certain volume, x cm³, of this solution was added to a 250 cm³ volumetric flask and the solution made up to the mark with water. What volume of the original $Na_2S_2O_3$ stock solution was taken if the new concentration is 0.100 M?

1.9 Volumetric analysis

Volumetric analysis involves the determination of the quantity or concentration of a substance by measuring volumes. The most common type of volumetric analysis determinations involves titrations. Titrations are usually either acid or base titrations, where the end point involves the complete neutralization of the acid (or base) by the other reagent, or redox titrations. At the end point of an acid/base titration the hydroxonium ions of the acid (H_3O^+, usually represented by H^+) have all reacted with the hydroxide ions of the base (OH^-). The end point is generally observed by the change in colour of an indicator.

The overall **ionic equation** for the neutralization is given by:

$$H^+ + OH^- \rightarrow H_2O$$

It is important to first balance the stoichiometric equation to determine the amount in moles of base (OH^-) that are required per mole of acid. This depends upon whether the acid is **mono-**, **di-** or **tribasic**.

In redox titrations, the oxidizing agent and reducing agent will react together in a stoichiometric ratio determined by the redox equation. The end point is reached when either one of the reagents has been used up, and is usually observed by the presence of excess of one of the reagents, or by an indicator that will change colour in the presence of the reagent in excess.

In any volumetric analysis problem involving titrations we can follow three steps:

Step 1: write a balanced stoichiometric equation;

Step 2: calculate the amount in moles of the reagent for which both the concentration and the volume are known;

Step 3: relate the amount in moles of each reagent using the equation and then calculate the number of moles of the other reagent.

The procedure can be seen in the following worked examples.

Worked example 1.9A

A 25.0 cm³ solution of sodium hydroxide was neutralized by 27.5 cm³ of aqueous sulfuric acid of concentration 0.250 mol dm⁻³. Calculate the concentration of the sodium hydroxide solution.

Solution

Step 1: Write the balanced equation:

$$H_2SO_4 + 2\, NaOH \rightarrow Na_2SO_4 + 2\, H_2O$$

> Check that you have the correct formula for the salt formed as this will determine the stoichiometry of the reaction.

From the balanced equation we can see that 1 mole of H_2SO_4 reacts with 2 moles of NaOH.

Step 2: Identify the reagent for which the concentration and volume are both known. In this case we are given these values for the sulfuric acid so we can find the amount in moles of acid from: $n_A = c_A \times V_A$.

Step 3: Relate the amount in moles of acid to the amount in moles of base and so determine the amount in moles of base.

$$n(\text{acid}) = 0.250 \text{ mol dm}^{-3} \times 27.5 \times 10^{-3} \text{ dm}^3 = 6.875 \times 10^{-3} \text{ mol}$$

Therefore $n(\text{base}) = 2 \times 6.875 \times 10^{-3} = 13.75 \times 10^{-3}$ mol

> We need 2 moles of base for each 1 mole of acid so multiply by 2

$$n_B = c_B \times V_B \text{ so, } c_B = \frac{n_B}{V_B} = \frac{13.75 \times 10^{-3} \text{ mol}}{25.0 \times 10^{-3} \text{ dm}^3} = 0.550 \text{ mol dm}^{-3}$$

Worked example 1.9B

A solution of sodium carbonate contains 5.30 g in 250 cm³ water. The solution was used to titrate HCl of concentration 0.250 mol dm⁻³. Calculate the volume of HCl solution required to neutralize 25.0 cm³ of the sodium carbonate solution.

Solution

This problem has an additional step as we aren't given the concentration of sodium carbonate directly but have to calculate it from the mass and molar mass of sodium carbonate and the volume of solution.

Sodium carbonate has the formula: Na_2CO_3.
The M_r of $Na_2CO_3 = 106$.

$$n(Na_2CO_3) = \frac{5.30 \text{ g}}{106 \text{ g mol}^{-1}} = 0.0500 \text{ mol.}$$

This amount in moles of Na_2CO_3 is dissolved in 250 cm³ water so the concentration is:

$$\frac{0.0500 \text{ mol}}{250 \times 10^{-3} \text{ dm}^3} = 2.00 \times 10^{-1} \text{ mol dm}^{-3} = 0.200 \text{ mol dm}^{-3}.$$

Step 1: We now turn to the titration and write the stoichiometric equation for the reaction:

$$2\, HCl + Na_2CO_3 \rightarrow 2\, NaCl + H_2O + CO_2$$

From the equation we can see that 2 moles of HCl are required for each mole of Na_2CO_3.

Step 2: The question asks us for the volume of HCl and we know both the volume and concentration of the Na_2CO_3. This step therefore involves finding the amount in moles of Na_2CO_3.

$$n(Na_2CO_3) = 25.0 \times 10^{-3} \text{ dm}^3 \times 0.200 \text{ mol dm}^{-3} = 5.00 \times 10^{-3} \text{ mol}$$

Step 3: Relate the amount in moles of acid to the amount in moles of base:

$$n(HCl) \text{ required} = 2 \times n(Na_2CO_3)$$

So, $n(HCl) = 2 \times 5.00 \times 10^{-3} = 10.00 \times 10^{-3}$ mol $= 0.0100$ mol

The concentration of the HCl is given as 0.250 mol dm⁻³ and so the volume required is:

$$V_{HCl} = \frac{n_{HCl}}{c_{HCl}} = \frac{0.0100 \text{ mol}}{0.250 \text{ mol dm}^{-3}} = 0.0400 \text{ dm}^3 = 40.0 \text{ cm}^3.$$

1.9 VOLUMETRIC ANALYSIS

Worked example 1.9C

Iron (II) sulfate is oxidized by potassium manganate(VII) in acid solution. The overall ionic equation is:

$$5\,Fe^{2+}(aq) + MnO_4^-(aq) + 8\,H^+(aq) \rightarrow Mn^{2+}(aq) + 5\,Fe^{3+}(aq) + 4\,H_2O(l)$$

The end point is indicated by the first permanent pink colour in the flask. What volume of 0.0150 M manganate (VII) solution will be required to just oxidize 25.0 cm³ 0.0200 M iron (II) sulfate solution?

Solution

Step 1: This is a redox titration. We are given the balanced overall equation in this case so from the equation we can see that 5 moles of Fe^{2+} are required for each mole of $KMnO_4$.

Step 2: The question tells us the volume and concentration of the Fe^{2+} solution in the flask so we can obtain the amount in moles of Fe^{2+}.

$$n(Fe^{2+}) = c_{Fe} \times V_{Fe} = 0.0200\,mol\,dm^{-3} \times 25.0 \times 10^{-3}\,dm^3 = 0.500 \times 10^{-3}\,mol$$

Step 3: $n(MnO_4^-) = \frac{1}{5}\,n(Fe^{2+})$ from the equation. So $n(MnO_4^-) = \frac{1}{5} \times (0.500 \times 10^{-3}\,mol) = 0.100 \times 10^{-3}\,mol$.

We are told that the concentration of $KMnO_4$ solution is 0.0150 M so we can use this information to get the volume of $KMnO_4$ from $V = \dfrac{n}{c} = \dfrac{0.100 \times 10^{-3}\,mol}{0.0150\,mol\,dm^{-3}} = 6.67 \times 10^{-3}\,dm^3 = 6.67\,cm^3$.

❓ Question 1.22

The percentage purity of a sample of aspirin prepared in the laboratory can be determined by 'back titration'. In an experiment 1.500 g sample of aspirin was dissolved in 50 cm³ NaOH of concentration 0.500 M. The solution was boiled to completely hydrolyse the aspirin according to the following equation:

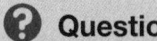

The resulting solution was titrated against a 0.500 M solution of HCl and 20.65 cm³ HCl were required to completely neutralize the excess sodium hydroxide. Calculate the percentage purity of the aspirin.
(Aspirin: $C_9H_8O_4$, $M_r = 180.15$)

❓ Question 1.23

Lead (II) nitrate solution, $(Pb(NO_3)_2)$ reacts with potassium iodide solution (KI) to precipitate lead (II) iodide.

(a) What volume of 0.100 mol dm⁻³ KI is required to react exactly with 25.0 cm³ $Pb(NO_3)_2$ of concentration 0.200 mol dm⁻³?

(b) This reaction can be used for the gravimetric determination of lead(II) ions in solution. What mass of PbI_2 would be precipitated from the lead(II) nitrate solution in (a)?

1.10 Limiting reagents and percentage yield

The **limiting reagent** in a reaction is the reactant that is used up first during a reaction. Once the limiting reagent is completely consumed, the reaction would cease to progress. The **theoretical yield** of a reaction is the amount (mass or moles) of product obtained when reaction ceases. This could be when the limiting reagent runs out or when the reaction reaches equilibrium and no more product can be formed under the existing conditions. The theoretical yield is the maximum yield of a product the reaction could give under ideal conditions.

The **percentage yield** is the actual yield in a reaction divided by the theoretical yield multiplied by one hundred to obtain a percentage. This can be calculated in terms of moles or mass of products.

$$\text{Percentage yield} = \frac{\text{actual yield}}{\text{theoretical yield}} \times 100\%$$

Worked example 1.10A

Calcium metal reacts in oxygen to form calcium oxide according to the equation:

$$2\,Ca\,(s) + O_2\,(g) \rightarrow 2\,CaO\,(s)$$

What is the maximum yield of calcium oxide that can be prepared from 4.20 g of Ca and 2.80 g of O_2?

If 4.52 g of CaO product are obtained calculate the percentage yield in the reaction.

Solution

The chemical equation for the reaction is provided for us. Before we can calculate the theoretical yield we first need to determine the limiting reagent. To do this we must find the amount in moles of each reactant and decide which is in excess. The reaction is therefore limited in yield by the reagent that has the fewer number of moles required by the chemical equation.

$$n\,(Ca) = \frac{4.20\,g}{40.1\,g\,mol^{-1}} = 0.105\,mol$$

$$n\,(O_2) = \frac{2.80\,g}{32.0\,g\,mol^{-1}} = 0.0875\,mol$$

From the chemical equation it can be seen that one mole of Ca requires half a mole of O_2. So 0.105 mol of Ca requires 0.052 mol of O_2 gas. As there are 0.0875 mol of O_2 gas available the O_2 is in excess and the metal is the limiting reagent.

The chemical equation tells us that 2 moles of Ca result in 2 moles of CaO product. 0.105 moles of Ca will give a maximum yield of 0.105 moles of CaO.

One mole of CaO has the mass 56.1 g.

Therefore 0.105 moles of CaO has the mass of 0.105 g × 56.1 g mol^{-1} = 5.89 g. This is the theoretical yield.

The reaction actually produces 4.52 g CaO. The percentage yield is therefore:

$$\frac{4.52\,g}{5.89\,g} \times 100\% = 76.7\%$$

Worked example 1.10B

The dehydration of ammonium ethanoate (ammonium acetate, $CH_3CO_2NH_4$) into ethanoic acid and ammonia is achieved by refluxing ammonium acetate with an excess of glacial ethanoic acid. In such a reaction 7.7 g ammonium acetate was found to produce 4.5 g dry ethanamide after fractional distillation. Calculate the percentage yield in the reaction.

Solution

In this reaction we are given the starting materials and the product but not the chemical equation. We are also told that the glacial ethanoic acid is in excess so we don't need to be concerned about the quantity of this involved. The first step in this case is to write a chemical equation:

$$CH_3CO_2NH_4 \rightarrow CH_3CONH_2 + H_2O$$

So this is a simple 1:1 reaction. If we can calculate the amount in moles of starting material we can obtain the theoretical yield. For this we need the molar mass of ammonium ethanoate, which is 77 g mol^{-1}.

$$n(CH_3CO_2NH_4) = \frac{7.7 \text{ g}}{77 \text{ g mol}^{-1}} = 0.10 \text{ mol}$$

Therefore theoretical yield = 0.10 mol ethanamide = 0.10 mol × 59 g mol^{-1} = 5.9 g.

The actual yield = 4.5 g and so the percentage yield = $\frac{4.5 \text{ g}}{5.9 \text{ g}} \times 100\% = 76\%$.

> We are told the starting material and the product and that this is a dehydration reaction and so water must be lost. The ethanoic acid is not required in the equation for the reaction.

 Question 1.24

Sodium borohydride (NaBH$_4$) is used industrially in many organic syntheses. It can be prepared by reacting sodium hydride with gaseous diborane (B$_2$H$_6$). Calculate the theoretical yield of NaBH$_4$ if 8.88 g of sodium hydride are allowed to react with 9.12 g of diborane. If 9.56 g NaBH$_4$ is obtained in the reaction calculate the percentage yield.

1.11 The gas laws

The physical behaviour of a gas can be described completely by four variables: pressure (p), volume (V), temperature (T), and amount in moles, (n). The variables are interdependent, which means that we can determine the value of one if we measure the other three.

Three relationships exist between the four variables. These are Boyle's Law, Charles's Law, and Avogadro's Law. These laws explain how the variables depend upon each other; you can read about them in your course textbooks. We will simply quote each law here and show how the laws are used to characterize the properties of a gaseous sample.

Boyle's Law

The pressure of a fixed amount of gas is inversely proportional to its volume at a given temperature. This can be expressed as:

$$p \propto \frac{1}{V}$$

Charles's Law

The volume of a fixed amount of gas is directly proportional to its temperature at constant pressure. This can be expressed as:

$$V \propto T$$

The combination of Boyle's Law and Charles's Law gives a relationship between pressure, volume, and temperature for a fixed mass of gas and can be written in the form:

$$p \propto \frac{T}{V}$$

This can be re-written in the form:

$$\frac{p_1 V_1}{T_1} = \frac{p_2 V_2}{T_2}$$

This allows corrections to be made for the change in volume of a gas between one set of pressure and temperature conditions and another, as well as other calculations.

Worked example 1.11A

If the volume of a fixed mass of argon gas at 1.00×10^5 Pa is 0.0312 m³ at 273 K calculate its volume at a pressure of 10×10^5 Pa and a temperature of 373 K.

Solution

In this question we have a sample of gas which has a certain volume under the initial conditions of temperature and pressure and we are required to calculate the new volume when both the pressure and temperature are increased. We can therefore use the combined equation that links pressure, volume, and temperature for a fixed amount in moles of a gas, i.e:

$$\frac{p_1 V_1}{T_1} = \frac{p_2 V_2}{T_2}$$

If we assume the information about the gas at 273 K represents the initial conditions (p_1, V_1, and T_1) then the second set of conditions at 373 K are represented by p_2, V_2, and T_2. The question asks for the new volume so the general equation must be rearranged to make V_2 the subject of the equation.

$$\frac{p_1 V_1}{T_1} = \frac{p_2 V_2}{T_2}$$

So, by rearranging: $V_2 = \frac{p_1 V_1 T_2}{T_1 p_2}$ and inserting values from the information given we get:

$$V_2 = \frac{1.00 \times 10^5 \text{ Pa} \times 0.0312 \text{ m}^3 \times 373 \text{ K}}{273 \text{ K} \times 10.00 \times 10^5 \text{ Pa}} = 4.26 \times 10^{-3} \text{ m}^3$$

> So, the combination of the increase in pressure by a factor of ten and the increase in temperature have resulted in a roughly ten times decrease in volume of the gas.

> Note that the units of pressure are Pa in both cases and temperature is in K so they cancel.

Ideal gas equation

The equation we have just seen that relates pressure, volume, and temperature:

$$p \propto \frac{T}{V}$$

can be written as $\frac{pV}{T} = k$. The constant in this equation, k, is the molar gas constant, R, and has the value 8.314 J K^{-1} mol^{-1}.

The **ideal gas law** states that for one mole of an ideal gas:

$$\frac{pV}{T} = R = 8.314 \text{ J K}^{-1} \text{mol}^{-1}$$

Table 1.3 The relationships between the commonly used units of pressure.

Unit	Symbol	Value
pascal	Pa	1 Pa = 1 N m^{-2}
bar	bar	1 bar = 1 × 10^5 Pa
torr	torr	1 torr = 1 mmHg = 133.32 Pa
atmosphere	atm	1 atm = 1.01325 Pa = 1.013 bar = 760 torr

For n moles of gas the equation can be written as: $\frac{pV}{nT} = R$. This is normally expressed as: $pV = nRT$.

Although we deal in practice with **real** gases it is often convenient to assume that all gases are **ideal** and so obey the ideal gas equation. Ideal gases are ones for which there are no intermolecular forces and in which the molecules have negligible volume. If we assume that all gases are ideal then the molar gas constant, R, has the same value for every gas and does not depend upon the nature of the molecule.

IUPAC (the International Union of Pure and Applied Chemistry) has defined standard pressure as 1.00×10^5 Pa (1 bar) and standard temperature as 273.15 K (although we generally use the value of 273 K).

> IUPAC has recently defined standard pressure as 10^5 Pa although 1 atm is commonly used still.

> Standard temperature of 273 K is different from the standard state temperature or standard ambient temperature of 298 K that is used in thermodynamics.

We saw earlier how there are various units of pressure. Table 1.3 shows the relationship between the commonly used units of pressure.

Worked example 1.11B

Calculate the volume of one mole of an ideal gas under standard conditions of temperature and pressure.

Solution

The ideal gas law allows us to calculate any one property of a sample of gas if we know the other conditions.

The ideal gas law can be expressed as: $pV = nRT$. Rearranging the equation to make the volume the subject gives us: $V = \frac{nRT}{p}$

Inserting the values for standard temperature and pressure and the gas constant allows us to calculate the value of the volume:

$$V = \frac{1 \text{ mol} \times 8.314 \text{ J K}^{-1} \text{ mol}^{-1} \times 273 \text{ K}}{1.00 \times 10^5 \text{ Pa}}$$

However, it is clear that we need to express pascals and joules in fundamental units in order to obtain a volume in m^3. From Table 1.3 we can see that 1 Pa = 1 N m^{-2} which in fundamental units is 1 kg m^{-1} s^{-2}. We also must convert J into fundamental units, which gives us 1 J = 1 kg m^2 s^{-2}.

The equation therefore becomes:

> Units of kg, K, mol, and s all cancel to leave a volume in m^3.

$$V = \frac{1 \text{ mol} \times 8.314 \text{ kg m}^2 \text{ s}^{-2} \text{ K}^{-1} \text{ mol}^{-1} \times 273 \text{ K}}{1.00 \times 10^5 \text{ kg m}^{-1} \text{ s}^{-2}} = 2269.7 \times 10^{-5} \text{ m}^3 = 0.0227 \text{ m}^3.$$

Converting from m^3 to dm^3 gives us: 0.0227×10^3 dm^3 = 22.7 dm^3.

We would obtain the same value no matter what gas we use. This volume, 22.7 dm^3, is therefore known as the molar volume of a gas under conditions of standard pressure (1×10^5 Pa) and temperature (273 K).

> → You may be more familiar with the value of 22.4 dm^3 as the molar volume, but this is measured at 1 atm pressure (101,325 Pa) and not 100,000 Pa.

The density of a gas

The previous calculation shows that one mole of any gas has approximately the same volume. The density of a gas therefore depends upon its molar mass. The greater the molar mass of the gas the higher the density. Density is the mass of a substance divided by its volume. The symbol used for density is d so: $d = \frac{m}{V}$. For gases the molar density $d = \frac{M}{V_m}$, where M is the molar mass of the gas and V_m is the volume occupied by one mole, which is a constant when measured at the same temperature and pressure. Thus the density of a gas depends only upon its molar mass, under the same conditions of temperature and pressure.

> The symbol (ρ) is sometimes used in place of d for density.

The ideal gas law can be rearranged to find the density of a gas. The ideal gas law is $pV = nRT$, which can be rearranged to $\frac{n}{V} = \frac{p}{RT}$, or the molar concentration.

The density of a gas is mass divided by volume $d = \frac{m}{V}$ where $m = nM$.

So $d = \frac{m}{V} = \frac{nM}{V} = \frac{p}{RT} M$.

This equation shows how the density of a gas varies with temperature and pressure.

Worked example 1.11C

Calculate the density of carbon dioxide under standard conditions of 1.00×10^5 Pa and 273 K.

Solution

Density is expressed by the mass of the gas divided by its volume. The ideal gas equation allows us to calculate the volume of a given amount in moles of a gas. The amount in moles of gas, n, can also be expressed as $\frac{m}{M}$. Therefore we can substitute for n in the ideal gas equation:

$$pV = nRT = \frac{mRT}{M}$$

To be able to calculate the density of a gas we need to be able to rearrange the equation to make $\frac{m}{V}$ the subject:

$$d = \frac{m}{V} = \frac{pM}{RT}$$

For any gas the molar mass is obtained from knowing the formula of the gas and the atomic masses. For the gas in the question, CO_2, the molar mass is 44.0 g mol^{-1}. We can therefore substitute the appropriate values into the equation for gas density:

$$d = \frac{1.00 \times 10^5 \text{ kg m}^{-1}\text{s}^{-2} \times 44.0 \times 10^{-3} \text{ kg mol}^{-1}}{8.314 \text{ J K}^{-1} \text{mol}^{-1} \times 273 \text{ K}} = 1.94 \text{ kg m}^{-3}$$

(or in g dm^{-3}) = 1.94 g dm^{-3}.

> Although the units of kg cancel on the top and bottom of the equation it is good practice to convert the molar mass to kg mol^{-1} in these types of calculations to keep the units consistent.

❓ Question 1.25

Calculate the density of molecular oxygen at 5.0 atm pressure and 27 °C.

❓ Question 1.26

The density of a gaseous compound was found to be 1.23 g dm^{-3} at 57 °C and 25.5 kPa. Calculate the molar mass of the compound.

Mixtures of gases

When we have a mixture of different gases in a container it behaves as a pure gas. Each gas in the container exerts its own pressure on the walls of the container. The pressure of each gas is directly proportional to the amount of gas present according to the ideal gas law. Gases in the same container occupy the same volume and experience the same temperature.

The pressure of each gas in a mixture is called the partial pressure of the gas. The partial pressure of a gas in a mixture is defined as the pressure the gas would exert if it occupied the container alone. The total pressure of all the gases in a mixture is equal to the sum of the partial pressures of the individual gases in the mixture. So for a mixture of three gases A, B, and C with partial pressures p_A, p_B, and p_C respectively the total pressure $p_T = p_A + p_B + p_C$. This relationship is known as **Dalton's Law of Partial Pressures**.

In order to calculate the total pressure of the gas we must be able to measure or calculate the partial pressures. The partial pressure of a gas is defined as the mole fraction of that gas in the mixture, X, multiplied by the total pressure of the gas, p_T. The mole fraction of a gas is the number of moles of the gas in the mixture divided by the total number of moles. So, if there are n_A moles of gas A in a mixture that contains a total of N moles of gases, the mole fraction of A is given by: $X_A = n_A/N$. The partial pressure of A, p_A is therefore given by $X_A \times p_T$.

Worked example 1.11D

In a sample of dry air weighing 1.00 g there was found to be 0.75 g nitrogen and 0.25 g oxygen. Calculate the partial pressures of nitrogen and oxygen in the sample if the overall pressure of the air is 1 atm.

Solution

From the analysis of the gas we know the mass of each component present in the mixture but not the amount in moles. The first step is therefore to convert the mass to moles by dividing by the molar mass of each gas.

$n(N_2) = 0.75 \text{ g}/28 \text{ g mol}^{-1} = 0.027$ mol.
$n(O_2) = 0.25 \text{ g}/32 \text{ g mol}^{-1} = 0.0078$ mol.

Therefore the total amount in moles of gas in the mixture, $N = 0.027 + 0.0078 = 0.0348$.
The partial pressure of each gas is given by $p_A = X_A \times p_T$.
Therefore the partial pressure of nitrogen, $p_{N_2} = (0.027/0.0348) \times 1 \text{ atm} = 0.78$ atm.
The partial pressure of oxygen, $p_{O_2} = (0.0078/0.0348) \times 1 \text{ atm} = 0.22$ atm

Question 1.27

The planet Mars has an atmosphere consisting mainly of carbon dioxide (95.32%) with nitrogen (2.75%) and argon (1.93%). The values given are percentage of each gas by mass in the atmosphere of Mars. The atmospheric pressure on Mars is roughly 600 Pa. Calculate the mole fractions of each gas and their partial pressures.

Question 1.28

What is the total pressure exerted by a mixture of 2.00 g of H_2 ($M_r = 2.016$), 10.00 g of N_2 ($M_r = 28.01$), and 12.0 g of Ar ($M_r = 39.95$) at 273 K in a 10.0 dm³ vessel (R = 0.08205 dm³ atm K⁻¹ mol⁻¹)?

Question 1.29

A sealed vessel contains equal masses of methane and carbon dioxide. The partial pressure of methane in the mixture is 0.350 atm. Calculate the partial pressure of carbon dioxide and the mole fraction of each gas.

Turn to the Synoptic questions section on page 170 to attempt questions that encourage you to draw on concepts and problem-solving strategies from several topics within a given chapter to come to a final answer.

Final answers to numerical questions appear at the end of the book, and full worked solutions appear on the book's website, where you can also find a set of bonus questions for each chapter. Go to www.oxfordtextbooks.co.uk/orc/chemworkbooks/.

2
Thermodynamics

2.1 Energy changes

Before embarking upon this topic there are a few terms that are useful to understand.

The system and surroundings

In a chemical reaction the **system** is the reaction of interest being studied and the **surroundings** are the rest of the space around the system. The system plus the surroundings is called the **universe**. There are different types of systems, which vary according to their transfer (or not) of **matter** and **energy**.

Matter can be relatively straightforward to understand in thermodynamics because we can see it, but energy is not as obvious. Energy can take the form of either heat or work:

- The transfer of energy as heat arises when there is a difference in temperature.
- Work is done **by** a reaction when it opposes a force, such as when a gas is produced and it expands against the external pressure. Work is done **on** the system if the atmosphere pushes in on the reaction and compresses any gaseous components.

An **open system** allows the exchange of both matter and energy between the system and the surroundings—for example, a beaker of boiling water. An example of a **closed system** is a stoppered reaction vessel where the exchange of matter between the system and its surroundings is not allowed but energy can be gained or lost as heat. An **isolated system** is one in which no exchange of matter or energy between the system and its surroundings is possible—for example, a vacuum flask with a tightly fitted lid.

Worked example 2.1A

Define the following systems as **open**, **closed**, or **isolated**.

(a) A sealed bottle of lemonade.

(b) A thermally insulated cool box.

(c) The earth.

Solutions

(a) This is a closed system as only energy can be exchanged between the system and its surroundings. Matter (the lemonade) cannot leave the system and enter the surroundings because the bottle is sealed.

(b) In theory this is an isolated system if the cool box is perfectly insulated.

(c) This is an open system as both matter and energy are constantly being both lost and gained by the system.

> **Question 2.1**
>
> Identify the following systems as **open**, **closed**, or **isolated**.
>
> (a) A reaction under reflux.
> (b) A polystyrene beaker of tea with a lid.
> (c) A torch battery.
> (d) A pressure cooker which is tightly sealed by its weight.

State and path functions and extensive and intensive properties

A **state function** is defined as a property of a system that depends only upon the current state of the system and not on the way of getting there. Some examples of state functions are mass, colour, pressure, and energy.

A **path function** is a property of a system that **does** depend upon the state of the system and the route taken to that state. Two important path functions are heat and work.

An **intensive** property doesn't depend upon the amount of the substance we have whereas an **extensive** property does. Dividing two extensive properties by each other gives an intensive property.

Worked example 2.1B

Define the following as **intensive** or **extensive** properties.

(a) Density.
(b) Volume.
(c) Concentration.
(d) Pressure.

Solution

(a) Density is an intensive property as it is constant no matter how much material we have. Density is equivalent to mass divided by volume (both extensive properties) and so is an intensive property.

(b) Volume is an extensive property as it increases with the amount of material—the more we have of a substance the bigger it gets.

(c) Concentration is an intensive property as it is constant throughout a homogeneous solution.

(d) Pressure of a gas is intensive as it doesn't depend upon the amount of gas. Wherever you measure the pressure of a fixed amount of gas in a container of fixed volume, its pressure will be the same. Pressure is equivalent to force divided by area; these are both extensive properties so their ratio gives an intensive property.

> **Question 2.2**
>
> Identify the following as **state** or **path** functions:
>
> (a) The enthalpy of formation of potassium chloride.
> (b) Temperature of a block of ice.
> (c) The volume of one mole of methane under standard conditions.
> (d) The heat contained in a beaker of water whose temperature is increased from 20 °C to 30 °C.

> **Question 2.3**
>
> Identify the following as **intensive** or **extensive** properties:
>
> (a) Temperature.
> (b) Energy.
> (c) Momentum.
> (d) Electrical charge.

2.2 Internal energy, *U*

The internal energy, **U**, of a system is the total energy contained in that system and is a state function. Internal energy is made up of all types of energy such as kinetic, vibrational, rotational, potential, etc.

In any reaction, energy may be changed from one form to another; it may also be used to do work, but it cannot be created or destroyed. Any energy lost by the system must be gained by the surroundings and vice versa.

One way of expressing the first law of thermodynamics is by using the following equation:

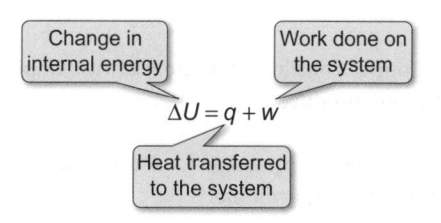

$$\Delta U = q + w$$

where ΔU is the change in internal energy, q is the heat transferred to the system, and w is the work done on the system.

→ The **First Law of Thermodynamics** states that energy can be neither created nor destroyed—only changed from one form to another.

→ When q is positive heat is transferred to the system and when w is positive work is done on the system.

This equation states that the change in internal energy, ΔU, in a closed system is the sum of the energy changes due to the transfer of heat, q, and work, w. In chemical reactions involving gases the work done is expansion work as the gas expands against the external pressure. In such cases the volume of the system increases, and the work done is given by the product of the external pressure, p_{ext} and the change in volume, ΔV:

Work = $w = -p_{ext} \Delta V$

If the reaction is carried out at constant volume $\Delta V = 0$, so the energy changes can only occur by transfer of heat:

$$\begin{aligned}\Delta U &= q + w \\ &= q - p_{ext} \Delta V \\ &= q - (p_{ext} \times 0) \\ &= q - 0 \\ &= q\end{aligned}$$

→ The negative sign for work implies energy is being **lost** by the system because when ΔV is positive the system is expanding and does work against the external pressure.

The transfer of heat at constant volume is represented by q_V and is the same as the internal energy change at constant volume.

Worked example 2.2A

A reaction is carried out in which 1.23 kJ of heat is released and 235 J of work is done on the surroundings. What is ΔU for the process?

Solution

We need to use the expression that relates the change in internal energy to the transfer of heat and work in a reaction: $\Delta U = q + w$.

In the following equation the appropriate values for q and w have been inserted into the expression for ΔU. Take care with both units and signs as shown in the equation.

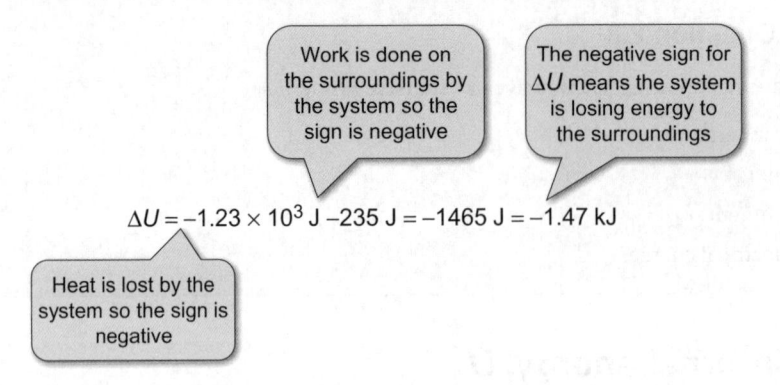

$$\Delta U = -1.23 \times 10^3 \text{ J} - 235 \text{ J} = -1465 \text{ J} = -1.47 \text{ kJ}$$

- Work is done on the surroundings by the system so the sign is negative
- The negative sign for ΔU means the system is losing energy to the surroundings
- Heat is lost by the system so the sign is negative

Worked example 2.2B

When a reaction was carried out at constant volume, 8.69 kJ of heat was absorbed; when run at constant pressure, 8.15 kJ of heat was absorbed. What is ΔU for the reaction and how much work is done when the reaction is run at constant pressure?

Solution

When a reaction is carried out at constant pressure it always implies that expansion work is being done by any gases that are produced.

Again, we use: $\Delta U = q + w$.

If the volume is constant there is no expansion work done on, or by, the system so $w = 0$. We can therefore determine ΔU at constant volume:

$$\Delta U = +8.69 \text{ kJ} + 0 = +8.69 \text{ kJ}$$

At constant pressure, work will be done by the system. Using $\Delta U = q + w$, $\Delta U = +8.69$ kJ and $q = +8.15$ kJ.

So, inserting values into the expression for ΔU:

$$+8.69 \text{ kJ} = +8.15 \text{ kJ} + w$$

and we rearrange the equation to make w the subject by subtracting 8.15 kJ from both sides to give:

$$w = +8.69 \text{ kJ} - 8.15 \text{ kJ} = +0.54 \text{ kJ}$$

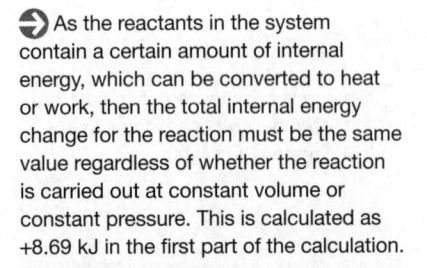

➔ As the reactants in the system contain a certain amount of internal energy, which can be converted to heat or work, then the total internal energy change for the reaction must be the same value regardless of whether the reaction is carried out at constant volume or constant pressure. This is calculated as +8.69 kJ in the first part of the calculation.

The value for q is positive as heat is absorbed by the system

➔ At constant pressure the system absorbs a smaller amount of heat (8.15 kJ). The difference between this and the overall internal energy change is 0.54 kJ, which is the amount of work done on the system. The positive sign for w tells us that the work is done on the system—for example, by compression.

❓ Question 2.4

During a reaction 2.87 kJ of heat is absorbed and the system does 445 J of work on the surroundings. Calculate the internal energy change for the reaction.

> **Question 2.5**
>
> A reaction run at constant volume releases 46.9 kJ of heat. When the same reaction is run at constant pressure it releases 45.8 kJ of heat. What is the work done on the system when the reaction is run at constant pressure?

Calculating expansion work done by a gas

Against a constant pressure

When a gas expands against a constant pressure such as in a gas syringe and the piston is pushed back against the external pressure or atmospheric pressure, as shown in Figure 2.1, the system is said to do work. The internal energy of the system decreases as it loses energy as work.

The total amount of work done can be calculated by knowing the external pressure (p_{ext}) and the total volume change:

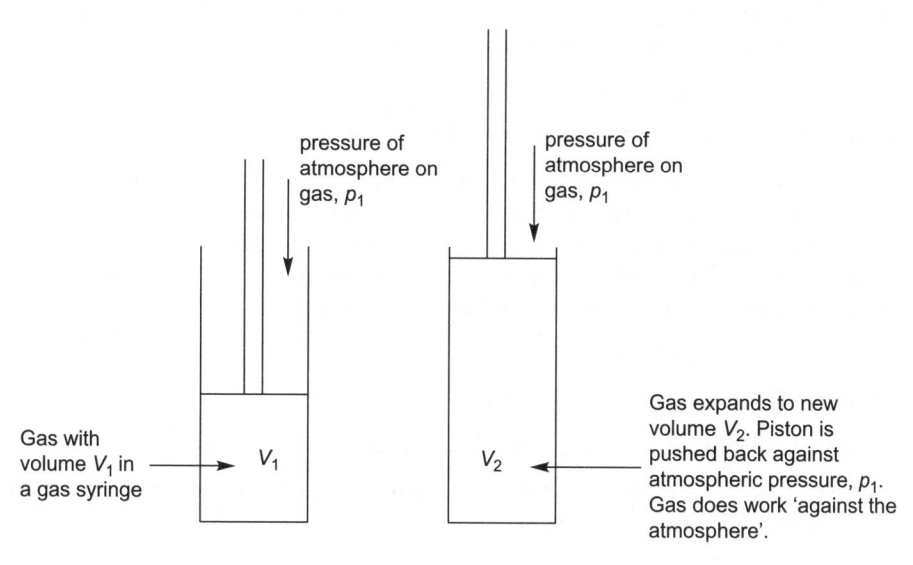

$$w = -p_{ext}\Delta V$$

Figure 2.1 Expansion of a gas at constant pressure in a gas syringe.

Against a changing external pressure— reversible isothermal expansion

In the previous examples the external pressure was constant. Reversible expansion occurs when a gas expands against a changing external pressure, and a reversible isothermal expansion occurs when the gas expands at a constant temperature against a changing external pressure. In an isothermal expansion the pressure of the gas falls as it expands; for the reaction to be reversible the external pressure also falls as the volume of the gas increases. Now the work done by the gas will change as the external pressure changes.

The infinitesimal work done at any time can be expressed as:

$$dw = -p_{ext}dV$$

> A change that can be reversed by an infinitesimal change in a variable (e.g pressure) is called a reversible change.

> This equation describes the infinitesimal amount of work done, dw, as the system expands through an infinitesimal change in volume, dV. At each stage the external pressure is equal to the current pressure of the gas, p.

> When the symbol Δ is replaced by d this implies an infinitesimally small change.

2 THERMODYNAMICS

where dV is the small change in volume against the external pressure at that time. In reversible expansion the external pressure is also changing as the gas expands, but the pressure is related to the volume by: $pV = nRT$. So, $p = \dfrac{nRT}{V}$.

Substituting for the pressure, p, the equation becomes:

$$dw = -\dfrac{nRTdV}{V}$$

and we can obtain the total work done by integrating as the volume changes from the initial to the final volume:

> We can take out the constants n, R, and T and integrate dV/V between the limits.

$$w = -\int_{V_{initial}}^{V_{final}} \dfrac{nRTdV}{V} = -nRT\int_{V_{initial}}^{V_{final}} \dfrac{dV}{V} = -nRT\ln\dfrac{V_{final}}{V_{initial}}$$

> The integral of dV/V is $\ln V$.

Worked example 2.2C

A gas expands by 750 cm^3 against a constant external pressure of 1.50 atm. Calculate the work done on, or by, the system.

Solution

We are told that the gas expands against an external pressure so we know that the system is doing work. If the external pressure is constant the work done can be calculated by $w = -p\Delta V$.

$$w = -1.50 \text{ atm} \times 750 \text{ cm}^3$$

Now we need to look at the units as we are looking for an answer in J.

- 1 atmosphere is equivalent to 101 325 Pa, which is the SI unit for pressure, so the pressure in atm should be multiplied by 101 325 Pa atm^{-1} to obtain a pressure in pascal.
- The volume change is given in cm^3 and so should be converted to m^3.
- There are 100 cm in 1 m and therefore there are 10^6 cm^3 in 1 m^3.
- We therefore have to multiply the volume by 10^{-6} to convert to m^3.

$$w = -1.50 \times 101\,325 \text{ Pa} \times 750 \times 10^{-6} \text{ m}^3$$

> Your calculator will give an answer to 9 significant figures but as we only have data accurate to 3 significant figures the answer given is −114 J

The derived units of Pa are kg m^{-1} s^{-2} which gives us:

$$w = -1.50 \times 101\,325 \text{ kg m}^{-1} \text{ s}^{-2} \times 750 \times 10^{-6} \text{ m}^3 = -113\,990\,625 \times 10^{-6} \text{ kg m}^2 \text{ s}^{-2} = -114 \text{ J}$$

Fortunately kg m^2 s^{-2} is the derived unit of the J so the answer becomes $w = -114$ J and the answer is given to three significant figures.

> ➔ A simpler way to deal with the units is to use the relationship that 1 Pa = 1 J m^{-3}. So we can convert the volume to m^3 and Pa to J m^{-3} and the m^3 will cancel:
>
> $w = -1.50 \times 101\,325$ Pa $\times 750 \times 10^{-6}$ m^3
> $= -1.50 \times 101\,325$ J m$^{-3} \times 750 \times 10^{-6}$ m^3
> $= -114$ J

Worked example 2.2D

A system of 5.00 moles of argon (assumed ideal) expands reversibly and isothermally at a temperature of $T = 298$ K from a volume of 50 L to 100 L. Calculate the work done by the system, the change in internal energy of the system, and the heat gained or lost by the system. ($R = 8.314$ J K^{-1} mol^{-1})

Solution

Reversible isothermal expansion implies that the external pressure is constantly changing as the gas expands and so we have to use the equation:

$$w = -nRT\ln\dfrac{V_{final}}{V_{initial}} \text{ to calculate the expansion work done.}$$

$$w = -5.00 \text{ mol} \times 8.314 \text{ J K}^{-1} \text{ mol}^{-1} \times 298.15 \text{ K} \times \ln\dfrac{100 \text{ L}}{50 \text{ L}}$$

$$= -12394.1 \text{ J} \times 0.693 = -8587 \text{ J} = -8.59 \text{ kJ}$$

The work done by the system is therefore 8.59 kJ.

As this expansion is isothermal the internal energy change of the system is zero.

Heat lost due to expansion is reversed by heat gain from the surroundings to maintain a constant temperature as the expansion is isothermal. In order to maintain the same temperature the gas must absorb 8.59 kJ heat.

> **Question 2.6**
>
> A sample of gas of volume 5.00 L is held in a piston at 300 K and a pressure of 2.00 atm. Calculate the work done on, or by, the system in each of the following processes.
>
> (a) The gas expands irreversibly against a constant external pressure of 1.00 atm to a final volume of 7.00 L.
> (b) Reversible isothermal expansion of the gas occurs to a final volume of 7.00 L.

> **Question 2.7**
>
> A sample of gas in a cylinder fitted with a piston of volume 3.00 dm^3 at 25 °C and 2.50 atm expands to 7.50 dm^3 by two different routes. Route 1 is an isothermal reversible expansion and Route 2 involves two steps. In the first step in Route 2 the gas is cooled at constant volume to 2.00 atm. In the second step the gas is heated and allowed to expand against a constant external pressure of 2.00 atm until the final volume is 7.50 dm^3. Calculate the work done in each path.

2.3 Enthalpy changes

The enthalpy change of a system, ΔH, is defined as the heat transferred between a reaction and the surroundings at constant pressure. It can be defined as:

$$\Delta H = \Delta U + p\Delta V$$

We know from the previous section that $\Delta U = q + w$ and $w = -p\Delta V$, so substituting ΔU for $q - p\Delta V$ in the equation for the enthalpy change ΔH gives:

$$\Delta H = q - p\Delta V + p\Delta V = q$$

So at constant pressure $\Delta H = q$.

→ At constant pressure a system involving gases can expand against the atmosphere. The expansion work is done by the system and this energy is lost by the system and is contained within the internal energy of the system, U. The accompanying heat in such a reaction is the enthalpy change, ΔH, of reaction. For most systems that don't involve gases ΔH is the same as q. For systems that do involve gases $\Delta H = q$ if the pressure is constant (which means the system can expand or contract).

Worked example 2.3A

RDX is an explosive whose chemical name is cyclotrimethylenetrinitramine and whose chemical formula is $C_3N_3(NO_2)_3H_6$. RDX detonates according to the equation:

$$C_3N_3(NO_2)_3H_6 \text{ (s)} \rightarrow 3\,N_2\text{ (g)} + 3\,H_2O\text{ (g)} + 3\,CO\text{ (g)}$$

The enthalpy change for the above reaction is approximately −1251 kJ. Assuming that all the gases given off behave as ideal gases and the reaction occurs at 298 K, calculate:

(a) The energy transferred as work when 1.0 mole of RDX detonates.
(b) The internal energy change when 1.0 mole of RDX detonates.

Solution

(a) We can calculate the work done to produce the gaseous products from the volume change and the pressure as shown previously.

$$w = -p\Delta V$$

We can assume all the gases behave as ideal gases and use the ideal gas equation:

$$pV = nRT$$

As we are looking for a change in volume multiplied by the pressure we introduce the symbol Δ, which means 'overall change'. If we add this to both sides of the equation we obtain:

$$\Delta pV = \Delta nRT$$

If both the temperature and pressure are constant then there is no overall change in these parameters. The only parameters in the equation that do change are the volume and the number of moles of gas on both sides of the equation. So we have ΔV and Δn, but p, R, and T are constant. The equation becomes:

$$p\Delta V = \Delta n_{gas}RT$$

As $w = -p\Delta V$ we can substitute for $p\Delta V$ so the equation becomes:

$$w = -\Delta n_{gas}RT$$

For the reaction in the example there are no moles of gas on the reactants side of the equation and nine moles formed on the product side. The change in number of moles, Δn, is therefore nine. So,

$$w = -9.0 \text{ mol} \times 8.314 \text{ J K}^{-1} \text{ mol}^{-1} \times 298 \text{ K} = -22\,298 \text{ J}$$
$$= -22 \text{ kJ}$$

The negative sign tells us that the reaction in which RDX detonates loses 22 kJ internal energy as work done in expansion caused by production of the gases.

(b) In calculating the internal energy change at constant pressure we use the relationship:

$$\Delta U = q + w$$

taking care to ensure the signs are correct.

For the detonation of RDX carried out at constant pressure, we have to remember that the enthalpy change of the reaction (ΔH) for one mole is equal to q, which we are told is −1251 kJ.

➔ As the enthalpy change is exothermic and work is done by the system both q and w are negative.

We can now substitute the values for q and w into the equation for the internal energy change:

$$\Delta U = q + w = -1251 \text{ kJ} + (-22 \text{ kJ}) = -1273 \text{ kJ}$$

Worked example 2.3B

During the decomposition of 0.001 mol hydrogen peroxide, how will the enthalpy change differ from the internal energy change when the reaction is carried out using a catalyst of MnO_2 at 1.00 atmosphere pressure and 273 K?

Solution

We know: $\Delta H = \Delta U + p\Delta V$.

So the difference between the enthalpy change (ΔH) and the internal energy change (ΔU) at constant pressure is given by rearranging $\Delta H = \Delta U + p\Delta V$ to give:

$$\Delta H - \Delta U = p\Delta V$$

First, we write a balanced equation for the reaction:

$$2\,H_2O_2\,(l) \rightarrow 2\,H_2O\,(l) + O_2\,(g)$$

Notice that the stoichiometry for this reaction is 2:2:1. The reaction uses 0.001 moles of hydrogen peroxide; the stoichiometry of the reaction therefore tells us that 0.0005 moles of oxygen gas are produced in total. This gas expands against the atmospheric pressure and does work as it expands. We can determine the work done by the system by calculating the product of the pressure and the volume change:

$$\Delta H - \Delta U = p\Delta V$$

$$p\Delta V = (\Delta n)RT$$

> State symbols here are very important. Hydrogen peroxide is a liquid and the product water is in the liquid state.

The initial number of moles of gas is zero and the final number 0.0005 so $\Delta n = +0.0005$. So,

$$p\Delta V = (\Delta n)RT = 0.0005\,\text{mol} \times 8.314\,\text{J mol}^{-1}\,\text{K}^{-1} \times 273\,\text{K}$$
$$= 1.135\,\text{J}$$

The difference between ΔH and ΔU is given by: $\Delta H - \Delta U = p\Delta V$.
So, $\Delta H - \Delta U = 1.135\,\text{J}$ for 0.0005 mol of O_2 gas produced.
So, for just one mole of gas the difference between ΔH and ΔU is:

$$\frac{1.135\,\text{J}}{0.0005\,\text{mol}} = 2270\,\text{J mol}^{-1} = 2.27\,\text{kJ mol}^{-1}$$

> **Question 2.8**
>
> The treatment of 0.050 mol sodium hydrogen carbonate with 0.10 M sulfuric acid leads to the production of carbon dioxide gas. Calculate the difference between the internal energy change for the reaction and the enthalpy change if the reaction is carried out at a constant pressure of 1.00 atmosphere and a temperature of 298 K.

2.4 Hess's Law

Hess's Law is another way of expressing the First Law of Thermodynamics. It can be stated as:

The total enthalpy change for a reaction is independent of the path by which the reaction occurs, provided the starting and finishing states are the same for each reaction path.

This law allows us to calculate energy changes for reactions that are difficult or impossible to measure experimentally.

The enthalpy change in a chemical reaction is represented by: $\Delta_r H^\ominus$
The $\ominus$ plimsoll symbol indicates that we are looking at the standard enthalpy change of reaction—that is, the change in enthalpy that occurs when the reactants are in their standard states. The **standard state** of a substance is the pure substance at 1 bar pressure.

> Standard conditions ($^\ominus$) refer to a pressure of 1 bar (1.00×10^5 Pa) and reactants and products in their standard states. Values of $\Delta_r H^\ominus$ are quoted at a specific temperature, usually 298 K.

Worked example 2.4A

Calculate the standard enthalpy of combustion of cyclopropane, C_3H_6, given the following data:

Enthalpy of formation of cyclopropane, $\Delta_f H^\ominus = +53.3\,\text{kJ mol}^{-1}$.
Enthalpy of combustion of carbon (graphite), $\Delta_c H^\ominus\,(C\,(gr)) = -393.5\,\text{kJ mol}^{-1}$.
Enthalpy of combustion of hydrogen, $\Delta_c H^\ominus\,(H_2\,(g)) = -285.8\,\text{kJ mol}^{-1}$.

> The standard enthalpy of formation of a substance, $\Delta_f H^\ominus$, is the enthalpy change when one mole of substance is formed from its elements in their standard states.

2 THERMODYNAMICS

> In a question of this type the aim is to construct an enthalpy triangle based on Hess's Law. If we can construct an enthalpy triangle, this allows us to equate the enthalpies required when going from one set of reactants to one set of products by two different routes—the basis of Hess's Law.

> Thermochemical equations can be treated in the same way as algebraic equations by cancelling unrequired reagents to leave the equation for the reaction of interest. It is important to keep track of signs and quantities:
>
> Write down the equation for the reaction of interest:
>
> $C_3H_6 (g) + 4½ O_2(g) \rightarrow 3 CO_2 (g) + 3 H_2O (l)$ $\Delta H = x$
>
> Write down the thermochemical equations for the known reactions:
>
> $3 C (gr) + 3 H_2 (g) \rightarrow C_3H_6 (g)$ $\Delta_c H^\ominus = +53.3$ kJ mol^{-1}
>
> $3 C (gr) + 3 O_2 (g) \rightarrow 3 CO_2 (g)$ $\Delta_c H^\ominus = -393 \times 3$ kJ mol^{-1}
>
> $3 H_2 (g) + ³⁄_2 O_2 (g) \rightarrow 3 H_2O (l)$ $\Delta_c H^\ominus = -285.8 \times 3$ kJ mol^{-1}
>
> Manipulate the three equations to give the equation of interest, cancelling 3 moles of C and 3 moles of H$_2$ on both sides:
>
> $C_3H_6 (g) + 3\cancel{C(gr)} + 3O_2(g) + 3\cancel{H_2(g)} + ³⁄_2 O_2 (g) \rightarrow 3\cancel{C(gr)} + 3\cancel{H_2(g)} + 3CO_2 (g) + 3 H_2O (l)$
>
> $C_3H_6 (g) + 4½ O_2 (g) \rightarrow 3 CO_2 (g) + 3 H_2O (l)$ $\Delta H = x$
>
> Inserting the values:
> -53.3 kJ mol^{-1} + $(-393.5 \times 3$ kJ mol$^{-1})$ + $(-285.8 \times 3$ kJ mol$^{-1}) = -2091$ kJ mol^{-1}

Solution

First, we write down the chemical reaction for which we want to calculate the enthalpy change.

$$C_3H_6 (g) + 4½ O_2 (g) \rightarrow 3 CO_2 (g) + 3 H_2O (l) \quad \Delta_c H^\ominus(C_3H_6 (g)) \tag{2.1}$$

$\Delta_c H^\ominus(C_3H_6(g))$ is the enthalpy change we want to calculate.

The reactants (C_3H_6 (g) + 4½ O_2 (g)) and the products (3 CO_2 (g) + 3 H_2O (l)) in equation 2.1 form two points on our Hess's Law triangle, as shown in Figure 2.2.

We are given the enthalpy of formation of cyclopropane in the question, so we add the elements carbon and hydrogen in their standard states as a third point on our Hess's Law triangle; the enthalpy of formation of cyclopropane forms the second side of the triangle in Figure. 2.2. The equation for this reaction is:

$$3 C (gr) + 3 H_2 (g) \rightarrow C_3H_6 (g) \quad \Delta_f H^\ominus(C_3H_6 (g)) \tag{2.2}$$

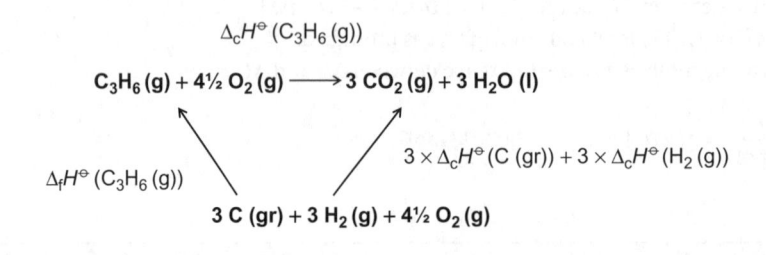

$$\Delta_f H^\ominus(C_3H_8(g)) + \Delta_c H^\ominus(C_3H_6(g)) = 3 \times \Delta_c H^\ominus(C(gr)) + 3 \times \Delta_c H^\ominus(H_2(g))$$

Figure 2.2 Construction of an enthalpy cycle for Worked example 2.4A.

We then complete the Hess's Law triangle by incorporating the information about the enthalpies of combustion of carbon and hydrogen. This forms the third side of our triangle.

$$C (gr) + O_2 (g) \rightarrow CO_2 (g) \quad \Delta_c H^\ominus(C (gr)) \tag{2.3}$$

$$H_2 (g) + ½ O_2 (g) \rightarrow H_2O (l) \quad \Delta_c H^\ominus(H_2 (g)) \tag{2.4}$$

In order to maintain the stoichiometry of the reaction, equation 2.3 and equation 2.4 must be multiplied by three as in the completed triangle shown in Figure 2.2.

We now have two routes starting from 3 moles of carbon and 3 moles of hydrogen to form 3 moles of CO_2 and 3 moles of H_2O. One route (equations 2.3 and 2.4) is directly by combustion and the second route is indirectly by first forming 1 mole of C_3H_6 (equation 2.2) and then by combustion of C_3H_6 to give 3 moles of CO_2 and 3 moles of H_2O (equation 2.1). By applying Hess's law we can equate the enthalpy changes for both routes to give:

$$\Delta_f H^\ominus (C_3H_6 (g)) + \Delta_c H^\ominus(C_3H_6 (g)) = 3 \times \Delta_c H^\ominus(C (gr)) + 3 \times \Delta_c H^\ominus(H_2 (g))$$

We then rearrange the equation to obtain an expression for $\Delta_c H^\ominus(C_3H_6)$, which is the enthalpy change we require:

$$\Delta_c H^\ominus(C_3H_6 (g)) = 3 \times \Delta_c H^\ominus(C (gr)) + 3 \times \Delta_c H^\ominus(H_2 (g)) - \Delta_f H^\ominus(C_3H_6 (g))$$

We can now insert the values as given in the question:

$$\Delta_c H^\ominus(C_3H_6 (g)) = (3 \times -393.5) \text{ kJ mol}^{-1} + (3 \times -285.8) \text{ kJ mol}^{-1} - 53.3 \text{ kJ mol}^{-1}$$
$$= -2091 \text{ kJ mol}^{-1}$$

> Drawing an enthalpy cycle, and ensuring the reaction arrows are in the correct direction and the signs of the enthalpy changes match the direction of enthalpy change, will help ensure your answer has the correct sign (exothermic or endothermic).

Worked example 2.4B

Calculate the standard enthalpy of formation of propane C_3H_8, $\Delta_f H^\ominus(C_3H_8(g))$ from the data below:

$H_2(g) + \frac{1}{2} O_2(g) \rightarrow H_2O(l)$ $\qquad \Delta_c H^\ominus(H_2(g)) = -285.8$ kJ mol^{-1}

$C(gr) + O_2(g) \rightarrow CO_2(g)$ $\qquad \Delta_c H^\ominus(C(gr)) = -393.5$ kJ mol^{-1}

$C_3H_8(g) + 5 O_2(g) \rightarrow 3 CO_2(g) + 4 H_2O(l)$ $\quad \Delta_c H^\ominus(C_3H_8(g)) = -2220$ kJ mol^{-1}

Solution

Again, we start by writing the balanced stoichiometric equation for the reaction whose enthalpy change we are asked to calculate. This forms the first side in our enthalpy triangle.

$3 C(gr) + 4 H_2(g) \rightarrow C_3H_8(g)$ $\qquad \Delta_f H^\ominus(C_3H_8(g))$

In this question we are given the enthalpies of combustion of the reactants on the left of the equation (i.e. $H_2(g)$ and $C(gr)$). We can use these values to construct a second side of the enthalpy triangle between $C(gr)$ and $H_2(g)$ and $CO_2(g)$ and $H_2O(l)$ as in Figure 2.3. As the formation of propane requires three moles of carbon and four moles of hydrogen we have to multiply the enthalpies of combustion by these values.

$$\begin{array}{c} \Delta_f H^\ominus(C_3H_8(g)) \\ 3\,C(gr) + 4\,H_2(g) \xrightarrow{} C_3H_8(g) + 5\,O_2(g) \\ \searrow \swarrow \\ 3 \times \Delta_c H^\ominus(C(g)) +   \Delta_c H^\ominus(C_3H_8(g)) \\ 4 \times \Delta_c H^\ominus(H_2(gr)) \\ 3\,CO_2(g) + 4\,H_2O(l) \end{array}$$

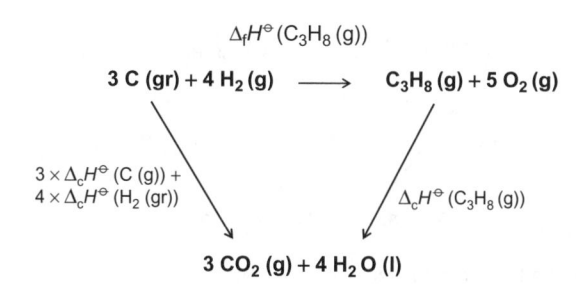

Figure 2.3 Construction of enthalpy triangle for Worked example 2.4B.

The third side of the triangle is the enthalpy of combustion of propane which we are also given in the question. Complete combustion of propane gives three moles of CO_2 and four moles of H_2O and requires five moles of $O_2(g)$ as shown in Figure 2.3.

Having drawn the complete enthalpy cycle we can use Hess's Law. Starting from the elements carbon and hydrogen and following the arrows in the anticlockwise and clockwise directions we can show that the enthalpy change produced by the total combustion of three moles of carbon and four moles of hydrogen must be the same as the enthalpy change for first forming one mole of propane from the reactants carbon and hydrogen $\Delta_f H^\ominus(C_3H_8(g))$ and then burning the propane in air $\Delta_c H^\ominus(C_3H_8(g))$:

$4 \times \Delta_c H^\ominus(H_2(g)) + 3 \times \Delta_c H^\ominus(C(gr)) = \Delta_f H^\ominus(C_3H_8(g)) + \Delta_c H^\ominus(C_3H_8(g))$

We now rearrange to get an expression for $\Delta_f H^\ominus$

$\Delta_f H^\ominus(C_3H_8(g)) = 4 \times \Delta_c H^\ominus(H_2(g)) + 3 \times \Delta_c H^\ominus(C(gr)) - \Delta_c H^\ominus(C_3H_8(g))$

We then insert the values:

$\Delta_f H^\ominus(C_3H_8(g)) = 4 \times -285.8$ kJ mol^{-1} $+ 3 \times -393.5$ kJ mol^{-1} $- (-2220$ kJ mol$^{-1})$

$= -1143.2$ kJ mol^{-1} -1180.5 kJ mol^{-1} $+ 2220$ kJ mol^{-1} $= -103.7$ kJ mol^{-1}

➔ Note that here we have the term $-\Delta_c H^\ominus(C_3H_8(g))$ in this equation, but the value of $\Delta_c H^\ominus(C_3H_8(g))$ itself is negative (-2220 kJ mol^{-1}) so we get a positive value in the next line when we carry out the calculation.

Worked example 2.4C

Given the following data:

$$H_2\,(g) + F_2\,(g) \rightarrow 2\,HF\,(g) \quad \Delta_f H^\ominus(HF\,(g)) = -537\ kJ\ mol^{-1}$$

$$C\,(s) + 2\,F_2\,(g) \rightarrow CF_4\,(g) \quad \Delta_f H^\ominus(CF_4\,(g)) = -680\ kJ\ mol^{-1}$$

$$2\,C\,(s) + 2\,H_2\,(g) \rightarrow C_2H_4\,(g) \quad \Delta_f H^\ominus(C_2H_4\,(g)) = +52.3\ kJ\ mol^{-1}$$

Calculate the enthalpy change for the fluorination of ethene:

$$C_2H_4\,(g) + 6\,F_2\,(g) \xrightarrow{\Delta_r H^\ominus} 2\,CF_4\,(g) + 4\,HF\,(g)$$

Solution

Again, we start by writing the balanced equation for the reaction we need the enthalpy change for. This forms the first side in the enthalpy triangle.

$$C_2H_4\,(g) + 6\,F_2\,(g) \xrightarrow{\Delta_r H^\ominus} 2\,CF_4\,(g) + 4\,HF\,(g)$$

In this problem the starting materials (carbon and hydrogen) are being fluorinated instead of oxygenated but the procedure is similar to that of the preceding examples. The triangle relating the enthalpy changes is shown in Figure 2.4.

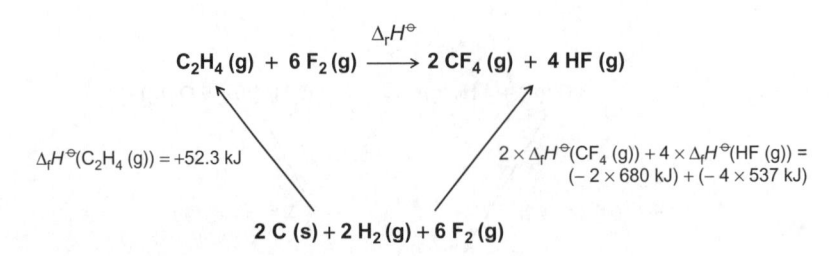

Figure 2.4 Enthalpy triangle for Worked example 2.4C.

Applying Hess's Law, starting from carbon, hydrogen, and fluorine:

$$\Delta_f H^\ominus(C_2H_4\,(g)) + \Delta_r H^\ominus = 2 \times \Delta_f H^\ominus(CF_4\,(g)) + 4 \times \Delta_f H^\ominus(HF\,(g))$$

$$+52.3\ kJ\ mol^{-1} + \Delta_r H^\ominus = 2 \times -680\ kJ\ mol^{-1} + 4 \times -537\ kJ\ mol^{-1}$$

Rearranging the equation to get an expression for the enthalpy of reaction:

$$\Delta_r H^\ominus = -1360\ kJ\ mol^{-1} - 2148\ kJ\ mol^{-1} - (+52.3\ kJ\ mol^{-1})$$

$$\Delta_r H^\ominus = -3560\ kJ\ mol^{-1} = -3.560\ MJ\ mol^{-1}$$

➡ To simplify the large value we have converted from kJ to MJ by dividing by 1000. Either answer would be satisfactory.

❓ Question 2.9

Given the following information, calculate the enthalpy of formation of propanone, C_3H_6O.

$$C_3H_6O\,(l) + 4\,O_2\,(g) \rightarrow 3\,CO_2\,(g) + 3\,H_2O\,(l) \quad \Delta_c H^\ominus(C_3H_6O\,(l)) = -1790\ kJ\ mol^{-1}$$

$$\Delta_f H^\ominus(CO_2\,(g)) = -393.5\ kJ\ mol^{-1}$$

$$\Delta_f H^\ominus(H_2O\,(l)) = -285.8\ kJ\ mol^{-1}$$

Question 2.10

Rocket engines can use methylhydrazine, CH_3NHNH_2, and dinitrogen tetroxide, N_2O_4, as a fuel. Given the following standard heats of formation, calculate the enthalpy change for the reaction:

$4\,CH_3NHNH_2\,(l) + 5\,N_2O_4\,(l) \rightarrow 4\,CO_2\,(g) + 12\,H_2O\,(l) + 9\,N_2\,(g)$

$\Delta_f H^\ominus(CH_3NHNH_2\,(l)) = +53\text{ kJ mol}^{-1}$ $\Delta_f H^\ominus(CO_2\,(g)) = -393\text{ kJ mol}^{-1}$

$\Delta_f H^\ominus(N_2O_4\,(l)) = +20\text{ kJ mol}^{-1}$ $\Delta_f H^\ominus(H_2O\,(l)) = -286\text{ kJ mol}^{-1}$

Question 2.11

The reaction of coke and steam produces a mixture called coal gas, which can be used as a fuel or as a starting material for other reactions. If we assume coke can be represented by graphite, the equation for the production of coal gas is:

$2\,C\,(gr) + 2\,H_2O\,(g) \rightarrow CH_4\,(g) + CO_2\,(g)$

Determine the standard enthalpy change for the reaction from the following standard enthalpies of reaction:

$C\,(gr) + H_2O\,(g) \rightarrow CO\,(g) + H_2\,(g) \quad \Delta_r H^\ominus = +131.3\text{ kJ}$ (2.11a)

$CO\,(g) + H_2O\,(g) \rightarrow CO_2\,(g) + H_2\,(g) \quad \Delta_r H^\ominus = -41.2\text{ kJ}$ (2.11b)

$CH_4\,(g) + H_2O\,(g) \rightarrow 3\,H_2\,(g) + CO\,(g) \quad \Delta_r H^\ominus = +206.1\text{ kJ}$ (2.11c)

2.5 The Born–Haber cycle

The Born–Haber cycle is an application of Hess's Law that allows us to calculate lattice enthalpies for an ionic solid. This is a closed path of steps representing different energy changes, and is illustrated in Figure 2.5. One of the steps is the formation of the solid lattice from the gaseous ions. The enthalpy change for this reaction is called the **lattice formation enthalpy**. For a solid with formula MX the lattice formation enthalpy is defined as the enthalpy change for the process $M^+\,(g) + X^-\,(g) \rightarrow MX\,(s)$ and an amount of heat equal to the lattice enthalpy is released when the solid forms from gaseous ions.

> Different textbooks define the lattice enthalpy in different directions. Here we are defining the lattice enthalpy as the enthalpy change associated with forming the lattice from its constituent ions in the gas state. This is sometimes termed the 'lattice formation enthalpy'. The important point is that creating the lattice from oppositely charged ions will always release energy and be exothermic, and separating the ions will always require energy and therefore be endothermic.

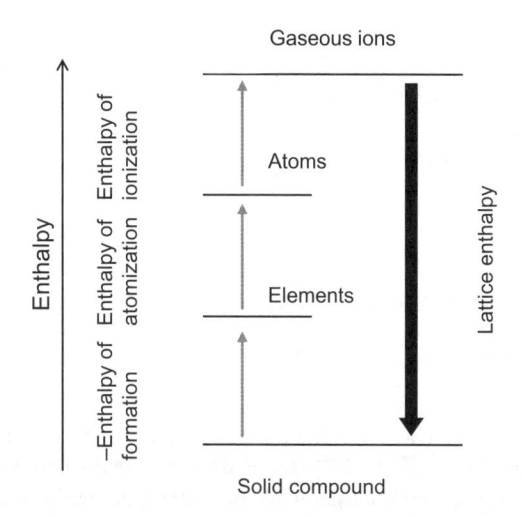

Figure 2.5 The Born–Haber cycle for the calculation of lattice enthalpy.

2 THERMODYNAMICS

The lattice enthalpy cannot be measured directly and so has to be calculated using other enthalpy changes that are measurable by experiment. These parameters include the standard enthalpy of formation of the substance, a value that can usually be measured experimentally, along with the energy changes to convert the elements into gaseous atoms and then to ionize them.

Worked example 2.5A

Use a Born–Haber cycle to find the lattice formation enthalpy for KCl given the following values:

$\Delta_f H^\ominus(KCl(s)) = -437 \text{ kJ mol}^{-1}$

$\Delta_a H^\ominus(K(s)) = 90 \text{ kJ mol}^{-1}$

$\Delta_a H^\ominus(Cl(g)) = 121 \text{ kJ mol}^{-1}$

$IE_1(K(g)) = 418.8 \text{ kJ mol}^{-1}$

$\Delta_{EA} H^\ominus(Cl(g)) = -349 \text{ kJ mol}^{-1}$

Solution

> Electron affinity $\Delta_{EA} H$ is the energy released as a gaseous atom of an element is attached to an electron:
> $X(g) + e^- \rightarrow X^-(g)$

Lattice formation enthalpy can be defined as the enthalpy change for the reaction:

$K^+(g) + Cl^-(g) \rightarrow KCl(s) \quad \Delta_{lat} H^\ominus$

The enthalpy cycle for the formation of KCl (s) is shown in Figure 2.6. Starting from metallic K and chlorine gas, Cl_2, there are two routes to the formation of KCl. One route involves the standard enthalpy of formation of KCl from the elements, $\Delta_f H^\ominus$. The second involves creating gaseous ions of K^+ and Cl^- from the elements, then allowing the ions to combine to form a solid lattice of KCl with a release in energy equivalent to the lattice enthalpy.

> $\Delta_a H$ represents the enthalpy change to convert the substance into one mole of gaseous atoms and IE_1 represents the first ionisation energy of the substance.

Application of Hess's Law gives us:

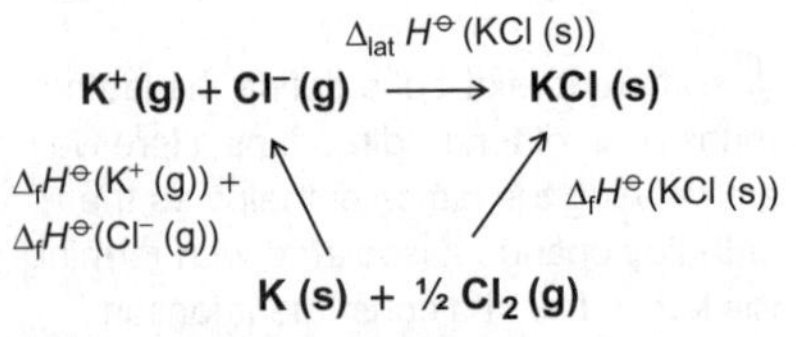

$\Delta_f H^\ominus(KCl(s)) = \Delta_f H^\ominus(K^+(g)) + \Delta_f H^\ominus(Cl^-(g)) + \Delta_{lat} H^\ominus(KCl(s))$

So: $\Delta_{lat} H^\ominus(KCl(s)) = \Delta_f H^\ominus(KCl(s)) - (\Delta_f H^\ominus(K^+(g)) + \Delta_f H^\ominus(Cl^-(g)))$

The quantities $\Delta_f H^\ominus(K^+(g))$ and $\Delta_f H^\ominus(Cl^-(g))$ represent the standard enthalpies of formation of the gaseous K^+ and Cl^- ions respectively from the elements.

For $K^+(g)$ this enthalpy change involves the enthalpy of atomization and the ionization energy:

$\Delta_f H(K^+(g)) = \Delta_{at} H^\ominus(K(s)) + IE_1(K(g))$

For the chloride ion this involves the enthalpy of atomization of chlorine gas (or half the bond enthalpy) plus the electron affinity:

$\Delta_f H^\ominus(Cl^-(g)) = \Delta_{at} H^\ominus(Cl(g)) + \Delta_{EA} H^\ominus(Cl(g))$

The overall Born–Haber cycle for the formation of KCl showing the direction of the enthalpy changes is shown in Figure 2.7.

We have already determined that:

$\Delta_{lat} H^\ominus(KCl(s)) = \Delta_f H^\ominus(KCl(s)) - (\Delta_f H^\ominus(K^+(g)) + \Delta_f H^\ominus(Cl^-(g)))$

$\Delta_{lat} H^\ominus(KCl(s)) = \Delta_f H^\ominus(KCl(s)) - (\Delta_{at} H^\ominus(K(s)) + IE_1(K(g)) + \Delta_{at} H^\ominus(Cl(g)) + \Delta_{EA} H^\ominus(Cl(g)))$

We can now insert the values provided into this equation.

$\Delta_{lat} H^\ominus = (-437 - (90 + 418.8 + 121 - 349)) \text{ kJ mol}^{-1}$
$= -718 \text{ kJ mol}^{-1}$

Figure 2.6 Hess's Law enthalpy triangle for calculating the lattice enthalpy of KCl (s).

The negative value of $\Delta_{lat} H^\ominus$ indicates that the direction of the enthalpy change used in the Born–Haber cycle, i.e. combining ions in the gaseous state to form the lattice, is exothermic.

You should note that in some data books and tables the lattice enthalpy relates to the endothermic process for separating the ions in a lattice to an infinite distance from each other and is therefore an endothermic process.

> It is important to think about the direction of enthalpy change and the sign when you quote or derive lattice enthalpies. Formation of the lattice from oppositely charged ions will almost always be exothermic and negative. Separation of the lattice into ions will be endothermic and positive.

2.5 THE BORN–HABER CYCLE

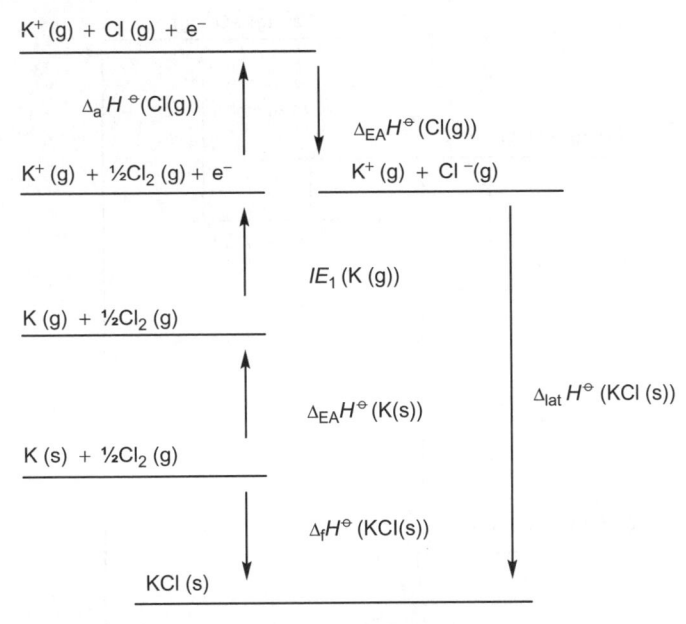

Figure 2.7 Born–Haber cycle for the formation of KCl.

❓ Question 2.12

(a) Identify and name each of the enthalpy terms on the diagram shown in Figure 2.8.

(b) Given the following data, calculate a value for ΔH_5:

$\Delta H_1 = +193$ kJ mol^{-1}, $\Delta H_2 = +590$ kJ mol^{-1}, $\Delta H_3 = +1150$ kJ mol^{-1},

$\Delta H_4 = +248$ kJ mol^{-1}, $\Delta H_6 = -3513$ kJ mol^{-1}, $\Delta H_7 = -635$ kJ mol^{-1}

(c) Use the value of ΔH_5 you have calculated to obtain the first electron affinity of oxygen, given that the second electron affinity of oxygen is +844 kJ mol^{-1}.

(d) Explain why the first electron affinity of oxygen is exothermic whilst the second electron affinity of oxygen is endothermic.

❓ Question 2.13

Given the thermodynamic data below, construct a Born–Haber cycle and use it to calculate the enthalpy of formation of ZnO.

Zn (s) → Zn (g) $\Delta_a H^\ominus = 130$ kJ mol^{-1}

½ O$_2$ (g) → O (g) $\Delta_a H^\ominus = 248$ kJ mol^{-1}

Zn (g) → Zn$^+$ (g) $IE_1 = 906$ kJ mol^{-1}

Zn$^+$ (g) → Zn^{2+} (g) $IE_2 = 1733$ kJ mol^{-1}

O (g) → O$^-$ (g) $\Delta_{EA} H^\ominus = -141$ kJ mol^{-1}

O$^-$ (g) → O^{2-} (g) $\Delta_{EA} H^\ominus = 780$ kJ mol^{-1}

Zn^{2+} (g) + O^{2-} (g) → ZnO (s) $\Delta_{lat} H^\ominus = -4002$ kJ mol^{-1}

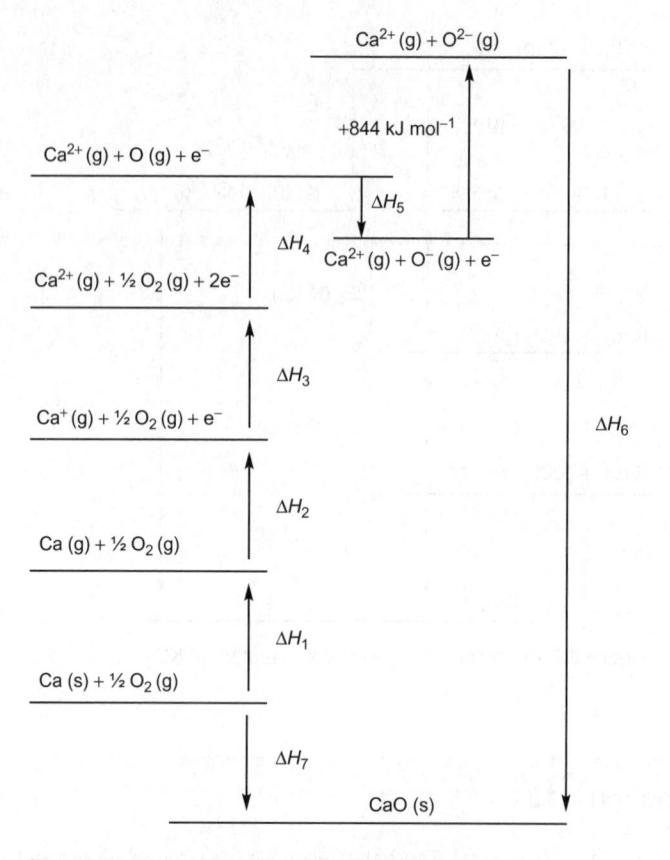

Figure 2.8 Born–Haber cycle for calcium oxide.

2.6 Bond enthalpies

An alternative method for calculating enthalpies of reaction is to use bond enthalpies. The **bond dissociation enthalpy** $D_{(A-B)}$ is defined as the enthalpy change when a bond A–B is broken under standard conditions in the gas phase. Because the strength of the same chemical bond (for example, C–C or C–H) depends upon the chemical environment of the bond, we generally use the **mean bond enthalpy**, $\overline{D}_{(A-B)}$.

In this type of problem the overall enthalpy change is given by the sum of the enthalpy terms for the bonds broken (endothermic) and the enthalpy released on bond formation (exothermic). When the bond is broken energy is required and ΔH is positive, when the bond is formed energy is released and ΔH is negative.

Worked example 2.6A

Estimate the standard enthalpy of reaction in which one mole of gaseous ethene, $C_2H_4(g)$, reacts with gaseous fluorine, $F_2(g)$, to form gaseous 1,2 difluoroethane, $C_2H_4F_2(g)$:

Mean bond enthalpies:

$\overline{D}_{(C=C)} = 612 \text{ kJ mol}^{-1}$

$\overline{D}_{(C-H)} = 412 \text{ kJ mol}^{-1}$

$\overline{D}_{(F-F)} = 158 \text{ kJ mol}^{-1}$

$\overline{D}_{(C-F)} = 484 \text{ kJ mol}^{-1}$

$\overline{D}_{(C-C)} = 348 \text{ kJ mol}^{-1}$

Solution

In order to solve this type of problem we need to calculate the total energy required to break the bonds in the reactant molecules, and the energy released when the new bonds in the product molecules are formed. The sum of these two quantities is the overall enthalpy change for the reaction which is the difference between the energy required and the energy released.

The equation involved is:

$\Delta_r H^\ominus$ = Total enthalpy required to break bonds + Total enthalpy released on bond formation

So, to find the enthalpy change required to break the bonds we must list the number and type of each bond that must be broken in the reactant molecules:

Bonds broken

$1 \times C=C = +612 \text{ kJ mol}^{-1}$

$4 \times C-H = 4 \times +412 \text{ kJ mol}^{-1} = +1648 \text{ kJ mol}^{-1}$

$1 \times F-F = +158 \text{ kJ mol}^{-1}$

Total = $(612 + 1648 + 158) \text{ kJ mol}^{-1} = +2418 \text{ kJ mol}^{-1}$

To find the enthalpy change as the new bonds are formed—an exothermic reaction:

Bonds formed

$2 \times C-F = -2 \times 484 \text{ kJ mol}^{-1} = -968 \text{ kJ mol}^{-1}$

$4 \times C-H = -4 \times 412 \text{ kJ mol}^{-1} = -1648 \text{ kJ mol}^{-1}$

$1 \times C-C = -348 \text{ kJ mol}^{-1}$

Total = $(-968 - 1648 - 348) \text{ kJ mol}^{-1} = -2964 \text{ kJ mol}^{-1}$

So, using the equation:

$\Delta_r H^\ominus$ = Total enthalpy required to break bonds + Total enthalpy released on bond formation

$= (+2418) + (-2964) \text{ kJ mol}^{-1}$

$= -546 \text{ kJ mol}^{-1}$

Note that these types of calculations are based on mean bond enthalpies and can only be used to give an average value for the answer. Consequently, the answer you obtain may not exactly match those quoted in books of thermochemical data. These values are usually obtained for specific molecules and not calculated using average bond enthalpies.

> Bond breaking always requires energy so is endothermic. By contrast, bond making releases energy as the atoms come together and so this is exothermic. It is important to remember this distinction.

> You may notice that there are four C–H bonds in the reactants and four C–H bonds in the products, so we could ignore the average bond enthalpy for the C–H bonds to simplify the calculation.

> **Question 2.14**
>
> Use the mean bond dissociation enthalpies given here to calculate the enthalpy of reaction for the complete combustion of propane in oxygen. Compare this value with the literature value of –2,220 kJ mol^{-1} and comment on the difference.
>
> | Bond type | $\overline{D}$/kJ mol^{-1} |
> |---|---|
> | C–H | 412 |
> | C–C | 348 |
> | O=O | 496 |
> | C=O | 805 |
> | H–O | 464 |

> **Question 2.15**
>
> Hydrazine, N_2H_4 (g), is often used as rocket fuel. It reacts exothermically with O_2 (g) forming gaseous products only.
>
> $$N_2H_4\,(g) + O_2\,(g) \rightarrow N_2\,(g) + 2\,H_2O\,(g)$$
>
> Use the values below to determine the enthalpy change for this reaction:
>
> $\overline{D}_{(N-N)} = 163$ kJ mol^{-1}
>
> $\overline{D}_{(N\equiv N)} = 945$ kJ mol^{-1}
>
> $\overline{D}_{(N-H)} = 390$ kJ mol^{-1}
>
> $\overline{D}_{(O=O)} = 496$ kJ mol^{-1}
>
> $\overline{D}_{(F-F)} = 158$ kJ mol^{-1}
>
> $\overline{D}_{(O-H)} = 464$ kJ mol^{-1}
>
> $\overline{D}_{(F-H)} = 562$ kJ mol^{-1}
>
> If fluorine gas was used in place of oxygen, determine whether the hydrazine/fluorine fuel mixture would be more or less efficient than with oxygen.

2.7 Heat capacity and calorimetry

The heat capacity, C, is the heat needed to raise the temperature of a substance by 1 K. The specific heat capacity, C_s, is the heat required to raise the temperature of 1 **gram** of a substance by 1 K.

Heat capacity can also be defined as the heat needed to raise the temperature of a substance by 1 °C. Although an actual temperature in kelvin (absolute temperature) is 273.15 degrees higher than the same temperature in Celsius, a change in temperature of one kelvin is the same as a change of one degree Celsius, as illustrated in Figure 2.9.

Therefore if an amount of heat, q, is transferred to a mass, m, of a substance and the temperature rises by ΔT, then the specific heat capacity is given by:

$$\text{Specific heat capacity} = \frac{\text{amount of heat supplied}}{\text{mass of substance} \times \text{temperature rise}}$$

This can be expressed as:

$$C_s = \frac{q}{m \times \Delta T} \text{ or } q = m \times C_s I \times \Delta T$$

2.7 HEAT CAPACITY AND CALORIMETRY

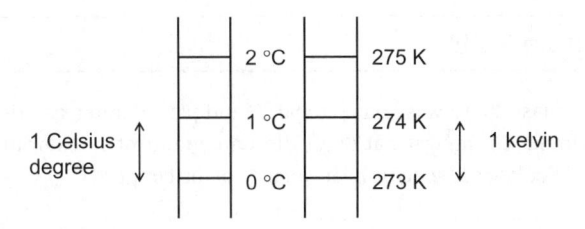

Figure 2.9 The Celsius and kelvin temperature scales. To convert from a temperature in degrees Celsius to kelvin you should add 273. However, the magnitude of a degree Celsius is the same as that of a kelvin—i.e. the temperature difference, ΔT, is the same.

If the heat released, q, is measured in joules (J), the mass in grams (g), and the temperature in kelvin (K) the units of specific heat capacity are given by $\frac{J}{g \times K} = J\,K^{-1}\,g^{-1}$. The specific heat capacity of water is 4.18 J K^{-1} g^{-1} or 4.18 J °C^{-1} g^{-1}.

Heat capacities can also be defined in terms of one **mole** of a substance. In this case the symbol is C_m and the expression given by:

$$C_m = \frac{q}{n \times \Delta T}$$

This expression can be rearranged to: $q = C_m \times n \times \Delta T$

When gases are involved the heat capacity of the gas depends upon whether measurements are carried out at constant volume or constant pressure. At constant volume the symbol used is C_v and at constant pressure C_p. These symbols actually refer to molar amounts of the gases, so n would be equal to one in the expression for C_v and C_p. Practically it is very difficult to measure C_v and so, often C_v is expressed in terms of C_p.

Worked example 2.7A

The specific heat capacity of water is 4.18 J g^{-1} °C^{-1}. Calculate the heat required to boil 500 cm^3 water at 20.0 °C ignoring the heat lost to the container.

Solution

Use the equation that relates heat energy to heat capacity, temperature change, and number of moles:

$$q = m \times C_s \times \Delta T$$

Note that the volume of water is given in cm^3 but specific heat capacity is expressed per gram of substance. The volume of water should therefore be converted to a mass of water using the density of water: 1 g cm^{-3}. So 500 cm^3 water is equivalent to 500 g water.

Now the values can be substituted in the equation:

$q = 500\text{ g} \times 4.18\text{ J °C}^{-1}\text{ g}^{-1} \times (100 - 20.0)\text{ °C} = 167\,200\text{ J} = 167\text{ kJ}$

In the previous question the heat capacity of the container was ignored but in most cases the reaction vessel will also absorb some of the heat thus requiring more energy to be provided. In reactions of this type the reaction vessel is typically a calorimeter and this must be first calibrated to determine its heat capacity. This is usually done electrically when the heat required is calculated from the current and voltage, and the temperature rise experienced by the calorimeter and known volume of water is equated to this.

→ The expression for q requires the temperature increase, ΔT, of the water. This is given by (100 − 20.0) °C = 80.0 °C simply because water boils at 100 °C.

Worked example 2.7B

A piece of copper of mass 30.0 g was heated to 90 °C and placed into a calorimeter of heat capacity 20 J °C^{-1} containing 100 cm^3 water at 20 °C. The temperature of the calorimeter and the water increased to 22 °C. Calculate the specific heat capacity of the copper. ($C_{s\,water} = 4.18$ J °C^{-1} g^{-1})

Solution

In this question the heat is provided by the copper and this is transferred to the water and the calorimeter. So we can write the equation:

Heat lost by copper = heat gained by water + heat gained by calorimeter.

The unknown is the specific heat capacity of the copper = $C_{s\,Cu}$
So the heat lost by the copper = $m_{Cu} \times C_{s\,Cu} \times \Delta T_{Cu}$
The heat gained by the water = $m_{water} \times C_{s\,water} \times \Delta T_{water}$
The heat gained by the calorimeter = $C_{cal} \times \Delta T_{cal}$

So, using the relationship that **the heat lost by copper = heat gained by water + heat gained by calorimeter**:

$$m_{Cu} \times C_{s\,Cu} \times \Delta T_{Cu} = m_{water} \times C_{s\,water} \times \Delta T_{water} + C_{cal} \times \Delta T_{cal}$$

Now we can place the values in the equation but note that the copper is originally at 90 °C but cools to 22 °C and the water and the calorimeter are initially at 20 °C but are heated to 22 °C by the heat lost from the copper:

$$30.0\text{ g} \times C_{s\,Cu} \times (90-20)\text{ °C} = 100\text{ g} \times 4.18\text{ J °C}^{-1}\text{ g}^{-1} \times (22-20)\text{ °C} + 20\text{ J °C}^{-1} \times (22-20)\text{ °C}$$

$$2100 \times C_{s\,Cu} \text{ g °C} = (418 + 20) \times 2 \text{ J}$$

$$C_{s\,Cu} = \frac{438 \times 2 \text{ J}}{2100 \text{ g °C}} = 0.42 \text{ J °C}^{-1}\text{ g}^{-1}$$

Worked example 2.7C

(a) In the calibration of a calorimeter, 80 kJ of heat were supplied electrically to the calorimeter which produced a temperature change of +8.4 °C. Calculate the heat capacity of the calorimeter.

(b) The calorimeter was used to investigate a certain combustion reaction. A temperature increase of +5.2 °C was produced. Calculate the heat, q, released in the reaction.

Solution

(a) Clearly in this question we can't ignore the heat capacity of the calorimeter as we are asked to calculate it in part (a) of the question. Calorimeters are normally calibrated electrically when the heat required to raise the temperature by a certain quantity can be determined from the voltage and current. Here we are given the electrical energy required and the temperature change so we can find the heat capacity, C, of the calorimeter directly.

$$q = C \times \Delta T$$

Rearranging to get an expression for C:

$$C = \frac{q}{\Delta T}$$

$$C = \frac{80 \text{ kJ}}{8.4 \,°\text{C}} = 9.5 \text{ kJ} \,°\text{C}^{-1}$$

(b) Once we have the heat capacity of the calorimeter we can use this to calculate the amount of heat required to produce any temperature change:

$$q = C \times \Delta T$$

$$q = 9.5 \text{ kJ} \,°\text{C}^{-1} \times 5.2 \,°\text{C} = 49 \text{ kJ}$$

> **Question 2.16**
>
> 1.00 g octane was burned in a calorimeter with a heat capacity of 994 J $°\text{C}^{-1}$. The temperature of the calorimeter rose by 4.88 °C. Calculate the heat released by the octane and the molar enthalpy of combustion of octane.

> **Question 2.17**
>
> A piece of iron of mass 200 g was heated from 20 °C to 100 °C and dropped in a calorimeter containing 300 ml water, originally at 20 °C. The temperature of the water and the calorimeter rose to 25 °C. Calculate the specific heat capacity of iron, $C_{s\,Fe}$, given that the heat capacity of the calorimeter = 30 J K^{-1}, and the specific heat capacity of water = 4.184 J $K^{-1} g^{-1}$.

Calorimetry is frequently used to determine fuel values and enthalpies of combustion as in Worked example 2.7D.

Worked example 2.7D

Biodiesel, obtained from plant material, is a possible alternative to petrol or diesel in internal combustion engines. Experiments have been carried out to determine the enthalpy of combustion of biodiesel and compare it with that of diesel. In one such experiment 5.00 cm^3 biodiesel was introduced into a calorimeter. The calorimeter was filled with 1 dm^3 water at 20.0 °C. The biodiesel was ignited and burned until completely oxidized. After correcting for heat losses a maximum temperature of 60.0 °C was recorded. Calculate the enthalpy of combustion of biodiesel and compare it with that of diesel = 44.8 MJ kg^{-1} and petrol = 44.4 MJ kg^{-1}. ($C_{s\,water}$ = 4.18 J $°\text{C}^{-1} g^{-1}$, density of biodiesel = 0.880 g cm^{-3})

Solution

The first task is to find the heat released by burning the biodiesel. This is obtained from the temperature increase of the water in the calorimeter. Once the total energy transferred has been determined this can be related to the quantity of biodiesel used.

Step 1: Use the equation which relates heat transferred to temperature rise of the water:

$$q = C_s \times m \times \Delta T$$

$$q = 4.18 \text{ J} \,°\text{C}^{-1} g^{-1} \times 1000 \text{ g} \times (60.0 - 20.0) \,°\text{C}$$

$$q = 167200 \text{ J} = 167 \text{ kJ}$$

→ Note here that heat is being transferred to the water and the volume has been converted to a mass in grams.

Step 2: Knowing the amount of heat released by 5.00 cm³ biodiesel, calculate the energy which would be released by 1 kg biodiesel and convert to obtain the correct units to make the comparison required in the question.

Density of biodiesel = 0.880 g cm⁻³

Volume of biodiesel = 5.00 cm³

Mass = density × volume = 0.880 g cm⁻³ × 5.00 cm³ = 4.40 g

So, 4.40 g biodiesel release 167 kJ on complete combustion. Therefore 1000 g (1 kg) would release: 167 kJ × 1000/4.40 = 38 000 kJ.

Converting to MJ we divide by 1000 = 38 000/1000 MJ = 38.0 MJ.

> 4.40 g biodiesel release 167 kJ, so
> 1.00 g release $\frac{167 \text{ kJ}}{4.40}$
> and 1000 g release $\frac{167 \text{ kJ} \times 1000}{4.40}$
> = 38 000 kJ

Note the conversion to MJ as 1MJ = 1 000 000 J = 1000 kJ.

CHECK: Comparing this with the enthalpy of combustion of diesel (44.8 MJ kg⁻¹) and petrol (44.4 MJ kg⁻¹) we can see this value is slightly lower but of the correct order of magnitude suggesting the calculation is correct.

❓ Question 2.18

In an equivalent experiment to determine the enthalpy of combustion of bioethanol, 2.50 g ethanol was added to a calorimeter. Complete combustion of this resulted in a temperature increase of 35.0 °C in the surrounding jacket which was filled with 500 cm³ water. If the heat capacity of the calorimeter was calibrated to be 50.0 J °C⁻¹ calculate the molar enthalpy of combustion of bioethanol. ($C_{s\ water}$ = 4.184 J °C⁻¹ g⁻¹, density of ethanol = 0.789 g cm⁻³)

❓ Question 2.19

In an experiment to determine the standard enthalpy of combustion of propan-1-ol a 0.60 g sample of the liquid was completely burnt in a calorimeter containing 0.50 dm³ water at 25.0 °C. The temperature of the calorimeter and its contents increased to 33.4 °C. The heat capacity of the calorimeter was determined independently by electrical measurements. It was found that 1.5 kJ electrical energy resulted in a 5.0 °C rise in temperature. Calculate the standard molar enthalpy of combustion of propan-1-ol. (Heat capacity of water = 4.184 J K⁻¹ g⁻¹)

2.8 Entropy and the Second Law of Thermodynamics

Entropy, S, can be defined in a number of ways. In chemistry, entropy is best considered as the number of ways of distributing the energy in a system between the molecules in that system. This definition leads to a concept of entropy which can be fairly easily understood and pictured. When the disorder of a system increases the entropy is said to increase. So, in general, the more energy there is in a system the more ways there are of distributing that energy. The more molecules there are in the system then the more ways there are of distributing the energy between the molecules. An increase in either of these factors can be said to increase the entropy of a system.

A more precise way of defining entropy is as a measure of the occupation of the different energy levels in a molecule. Whichever definition or explanation is used, however, entropy is a measure of the degree of disorder or randomness in a system.

Entropy is a state function and the change in entropy is given by the difference between the final and initial entropies of a system:

$\Delta S = S_{(final)} - S_{(initial)}$

Worked example 2.8A

Comment on the direction of entropy change, ΔS, in the following transformations:

(a) Evaporation of water from a boiling kettle.
(b) $2 H_2 (g) + O_2 (g) \rightarrow 2 H_2O (g)$
(c) Crystallization of solid sodium chloride from molten salt.
(d) $2 NaNO_3 (s) \rightarrow 2 NaNO_2 (s) + O_2 (g)$

Solution

(a) As liquid water changes to steam the gaseous molecules are more dispersed and are therefore more disordered than in the liquid state and therefore ΔS is positive.

(b) This equation represents the explosion of hydrogen and oxygen gas to give water in the vapour state. This is a homogeneous (single phase) reaction but there are fewer numbers of moles of gas on the right-hand side of the equation than the left. This reduction in number of moles is responsible for a decrease in entropy, so ΔS is negative.

(c) Here an ordered crystalline lattice is being formed from ions in the liquid state and so the entropy decreases, ΔS is negative.

(d) Decomposition of two moles of solid sodium nitrate produces two moles of sodium nitrite, but also one mole of gaseous oxygen. So overall the entropy increases, ΔS is positive.

> **Question 2.20**
>
> Comment on the direction of entropy change, ΔS, in the following transformations:
>
> (a) $C_6H_{12}O_6 (s) \rightarrow 2 C_2H_5OH (l) + 2 CO_2 (g)$
> (b) Dissolving sugar in water.
> (c) $N_2 (g) + 3 H_2 (g) \rightarrow 2 NH_3 (g)$
> (d) $[Co(NH_3)_6]^{3+} (aq) + 3 NH_2CH_2CH_2NH_2 (aq) \rightarrow$
> $[Co(NH_2CH_2CH_2NH_2)_3]^{3+} (aq) + 6 NH_3 (aq)$

Determining entropy change quantitatively

The Second Law of Thermodynamics states that:

Spontaneous processes are those that increase the total entropy of the universe.

This means that for a process to be spontaneous there is an associated increase in total entropy of the system and the surroundings.

For a reversible change at constant temperature the change in entropy is equal to heat absorbed or evolved divided by the constant temperature in kelvin.

Thus:

$$\Delta S = \frac{q_{rev}}{T}$$

Worked example 2.8B

Calculate the change in entropy when a large block of ice loses 50 J of heat at 0.0 °C in a freezer. Comment on the sign of the entropy change.

Solution

Using the relationship $\Delta S = \dfrac{q_{rev}}{T}$ we can insert the values given in the question to derive a value for the entropy change:

$$\Delta S = -\dfrac{50\,J}{273\,K} = -0.18\,J\,K^{-1}$$

→ The temperature must be changed to kelvin by adding 273 as the units of entropy are J K^{-1}.

→ The heat energy is being lost from the system and so the direction of entropy change is negative.

Think about the answer: the overall entropy change is negative. This is consistent with energy being lost by the system as the ice cools.

Question 2.21

Calculate the entropy change when one mole of benzene vapourizes at its boiling point under atmospheric pressure and comment on the sign. ($\Delta_{vap}H^\circ = 30.8$ kJ mol^{-1}, $T_b = 80.1\,°C$)

Change of entropy with temperature

In a reversible chemical reaction carried out at constant pressure we can equate the heat, q_{rev}, to the molar heat capacity at constant pressure, C_p, and the temperature change using the expression for any number of moles, n, of substance:

$$q_{rev} = n \times C_p \times \Delta T$$

→ The heat transferred at constant pressure ($q_{rev} = \Delta H$) is the same as the enthalpy change.

For 1 mole of substance: $q_{rev} = C_p \times \Delta T$.

The entropy change of 1 mole of substance between two temperatures T_i (initial temperature) and T_f (final temperature) can be derived to be:

$$\Delta S = C_p \ln\dfrac{T_f}{T_i}$$

→ Molar entropy increases with temperature as the ln term will always be positive provided $T_f > T_i$, although the rate of change is not linear as this equation suggests.

Worked example 2.8C

Calculate the entropy change on heating 1 mole of argon from 298 K to 500 K. ($C_{p\,Ar} = 20.8$ J K^{-1} mol^{-1})

Solution

The equation that relates entropy change and temperature is:

$$\Delta S = C_p \ln\dfrac{T_f}{T_i}$$

The entropy change is found by substituting values into the equation for the heat capacity of argon and the initial and final temperatures.

$$\Delta S = 20.8\,J\,K^{-1}\,mol^{-1} \times \ln\dfrac{500\,K}{298\,K} = 10.8\,J\,K^{-1}\,mol^{-1}$$

Question 2.22

Calculate the change in entropy ΔS when 1 mole of an ideal gas at a temperature of 298 K is heated at constant pressure until the temperature reaches 423 K. The heat capacity of the gas is $C_p = 12.48$ J K^{-1} mol^{-1}.

Standard reaction entropies

To calculate the change in entropy in a reaction we need to know the molar entropies of all the substances involved. The change in entropy can then be determined by subtracting the molar entropies of the reactants from the molar entropies of the products.

$$\Delta_r S^\ominus = \Sigma v_i S_m^\ominus(\text{products}) - \Sigma v_i S_m^\ominus(\text{reactants})$$

Where:

- $\Sigma v_i S_m^\ominus(\text{products})$ is the standard molar entropy of all the products
- $\Sigma v_i S_m^\ominus(\text{reactants})$ is the standard molar entropy of all the reactants.

The molar entropies of the products and reactants are multiplied by v, the stoichiometric coefficient or the amount in moles of each reagent in the stoichiometric equation.

→ The stoichiometric coefficient, (v_i), represents the amount in moles of i in the balanced chemical equation. It has no units.

Worked example 2.8D

Calculate the standard entropy change for the following reaction given the standard molar entropies below and comment on the sign of the change:

$$CH_4(g) + H_2O(g) \rightarrow CO(g) + 3\,H_2(g)$$

Substance	$S_m^\ominus / \text{J K}^{-1}\,\text{mol}^{-1}$
CH_4 (g)	186
H_2O (g)	189
CO (g)	198
H_2 (g)	131

Solution

The equation to use here is: $\Delta_r S^\ominus = \Sigma v S_m^\ominus(\text{products}) - \Sigma v S_m^\ominus(\text{reactants})$

Substituting in the entropies of the reactants and products, and remembering to multiply by the stoichiometric coefficients we obtain:

$$\Delta_r S^\ominus = (198 + 131 \times 3)\,\text{J K}^{-1}\,\text{mol}^{-1} - (186 + 189)\,\text{J K}^{-1}\,\text{mol}^{-1}$$

$$\Delta_r S^\ominus = +216\,\text{J K}^{-1}\,\text{mol}^{-1}$$

Here we multiply by three because there are three moles of H_2 in the equation.

→ Here we see an increase in entropy as there are more moles of gaseous product molecules. There is therefore greater disorder within the products compared to the reactants.

> **? Question 2.23**
>
> Calculate the standard entropy change for the reaction between ammonia and hydrogen chloride given the standard molar entropies below and comment on the sign of the change:
>
> $$NH_3(g) + HCl(g) \rightarrow NH_4Cl(s)$$
>
> | Substance | $S_m^\ominus / \text{J K}^{-1}\,\text{mol}^{-1}$ |
> |---|---|
> | NH_3 (g) | 193 |
> | HCl (g) | 187 |
> | NH_4Cl (s) | 94.6 |

Question 2.24

Calculate the standard entropy change for the reaction between phosphorus and chlorine to give phosphorus trichloride given the standard molar entropies below and comment on the sign of the change:

$$P_4(g) + 6\,Cl_2(g) \rightarrow 4\,PCl_3(g)$$

Substance	$S_m^\ominus$/J K^{-1} mol^{-1}
$P_4(s)$	41
$Cl_2(g)$	223
$PCl_3(g)$	312

Question 2.25

The standard enthalpy of formation of liquid ethanol, C_2H_5OH (l), is -278 kJ mol^{-1}, and the enthalpy change for the reaction below is $+112$ kJ mol^{-1}.

$$C_2H_5OH\,(l) \rightarrow CH_3CHO\,(g) + H_2\,(g)$$

(a) Calculate the standard enthalpy of formation of ethanal in the gas phase, CH_3CHO (g).
(b) With explanations, predict the **sign** of the entropy change, $\Delta_r S^\ominus$, for this reaction.
(c) Using the following values for $S_m^\ominus$ determine the entropy change in the reaction.

Substance	$S_m^\ominus$/J K^{-1} mol^{-1}
C_2H_5OH (l)	161
CH_3CHO (g)	266
H_2 (g)	131

Entropy changes in the surroundings

The Second Law of Thermodynamics states that for a reaction to be spontaneous the total entropy of the universe must increase. The entropy of the universe refers to the entropy of the system plus the surroundings: $\Delta S_{tot} = \Delta S + \Delta S_{surr}$

In the previous section we saw how the entropy of the system, ΔS, is calculated, but a change in entropy in the system will typically involve a change in entropy of the surroundings, ΔS_{surr} (as heat is often transferred). Assuming that the surroundings are so large that their temperature and pressure do not change significantly, then for any heat transfer at constant pressure the heat transferred is equal to the enthalpy change: $q_{surr} = -\Delta H$.

So, the equation $\Delta S_{surr} = \dfrac{q_{surr}}{T}$ becomes $\Delta S_{surr} = -\dfrac{\Delta H}{T}$

Therefore to confidently state whether a reaction is spontaneous we need to be aware of the entropy change of the system and the surroundings.

Worked example 2.8E

State whether the formation of hydrogen chloride from its elements in their most stable form is spontaneous at 25 °C.

$$H_2(g) + Cl_2(g) \rightarrow 2\,HCl(g) \quad \Delta_r H^\ominus = -185\text{ kJ mol}^{-1}$$

Substance	$S_m^\ominus$/J K^{-1} mol^{-1}
H$_2$ (g)	131
Cl$_2$ (g)	223
HCl (g)	187

Solution

There are three stages to complete in this question. We first need to find the entropy change of the system using the stoichiometric equation and the relationship that:

$$\Delta S^\ominus = \Sigma v S_m^\ominus(\text{products}) - \Sigma v S_m^\ominus(\text{reactants})$$

We then need to calculate the entropy change of the surroundings using the enthalpy change of reaction and $\Delta S_{\text{surr}} = -\dfrac{\Delta H}{T}$.

Then we can sum the entropy change of the system and surroundings to obtain the overall entropy change.

So, to obtain the entropy change of the system:

$$\Delta S^\ominus = 2 \times 187 \text{ J K}^{-1}\text{ mol}^{-1} - (131 + 223) \text{ J K}^{-1}\text{ mol}^{-1}$$

$$= (374 - 354) \text{ J K}^{-1}\text{ mol}^{-1}$$

$$= 20 \text{ J K}^{-1}\text{ mol}^{-1}$$

To obtain the entropy change of the surroundings we substitute the enthalpy change and the temperature into the equation:

$$\Delta S_{\text{surr}} = -\frac{\Delta H}{T} = -\frac{-185 \times 10^3 \text{ J mol}^{-1}}{298 \text{ K}} = +621 \text{ J K}^{-1}\text{ mol}^{-1}$$

So, we can now find the total entropy change:

$$\Delta S_{\text{tot}} = \Delta S + \Delta S_{\text{surr}}$$

$$= +20 \text{ J K}^{-1}\text{ mol}^{-1} + 621 \text{ J K}^{-1}\text{ mol}^{-1}$$

$$= 641 \text{ J K}^{-1}\text{mol}^{-1}$$

The total entropy change of the universe is therefore positive so the reaction will be spontaneous.

> **Question 2.26**
>
> Calculate whether the formation of cyclohexane from its elements in their standard states at 25 °C is spontaneous given the following data:
>
> | Substance | $S_m^\ominus$/J K^{-1} mol^{-1} |
> |---|---|
> | C (gr) | 5.7 |
> | H$_2$ (g) | 131 |
> | C$_6$H$_{12}$ (l) | 204 |
>
> $\Delta_f H^\circ(\text{C}_6\text{H}_{12}(\text{l})) = -156 \text{ kJ mol}^{-1}$

Gibbs free energy

The Gibbs free energy change for a process that occurs at constant temperature is given by:

$$\Delta G = \Delta H - T\Delta S$$

The expression is derived in standard textbooks from the Second Law of Thermodynamics. The Gibbs free energy change combines the enthalpy change and the entropy change in a reaction and allows us to predict whether the reaction will be spontaneous at constant temperature and pressure.

From the Second Law of Thermodynamics, the requirement for a reaction to be spontaneous is that $\Delta S_{(total)} > 0$. Therefore at constant temperature and pressure the direction of spontaneous change must be towards lower free energy and so $\Delta G_{(p,T)} < 0$.

Worked example 2.8F

Decide whether, given the conditions quoted, the following reactions will be spontaneous:
(a) Enthalpy change, ΔH, is exothermic, entropy, ΔS, increases.
(b) Enthalpy change, ΔH, is endothermic, entropy, ΔS, increases but $T\Delta S < \Delta H$.
(c) Enthalpy change, ΔH, is endothermic, entropy, ΔS, decreases.

Solution

(a) ΔH is negative, $(T\Delta S)$ is positive so, in $\Delta G = \Delta H - T\Delta S$, ΔG must always be overall negative and so this reaction is always spontaneous.
(b) ΔH is positive, $(T\Delta S)$ is positive so in $\Delta G = \Delta H - T\Delta S$, $T\Delta S < \Delta H$ so ΔG is overall positive and reaction not spontaneous.
(c) ΔH is positive and $(T\Delta S)$ is negative so in $\Delta G = \Delta H - T\Delta S$, $(-T\Delta S)$ is positive so ΔG is overall positive and reaction not spontaneous.

Worked example 2.8G

(a) What values of ΔH would permit spontaneous reaction if there was a decrease in entropy of the system?
(b) If a reaction is spontaneous and endothermic, what can be said about the change in entropy?
(c) Estimate the temperature at which it is thermodynamically possible for carbon to reduce iron(III) oxide by the endothermic reaction:

$$2\,Fe_2O_3\,(s) + 3\,C\,(s) \rightarrow 4\,Fe\,(s) + 3\,CO_2\,(g)\quad \Delta_r H^\ominus = +467.9\text{ kJ mol}^{-1},\ \Delta_r S^\ominus = 558.3\text{ J K}^{-1}\text{mol}^{-1}$$

Solution

(a) If there is a decrease in entropy then $(T\Delta S)$ will be negative, but this results in $(-T\Delta S)$ being positive. So for ΔG to be overall negative the value of ΔH must be more negative than that of $T\Delta S$.
(b) ΔH is positive hence for ΔG to be negative $T\Delta S > \Delta H$ so ΔS must be positive.
(c) Using the relationship $\Delta G = \Delta H - T\Delta S$, ΔG changes sign when $\Delta H = T\Delta S$.

We can set ΔG as equal to zero so:

$$T = \Delta H/\Delta S = 467.9 \times 10^3\text{ J mol}^{-1}/558.3\text{ J mol}^{-1}\text{ K}^{-1} = 838.1\text{ K}$$

Worked example 2.8H

Calculate the standard Gibbs energy change at 298 K for the following reaction using the data below:

$$CH_4\,(g) + N_2\,(g) \rightarrow HCN\,(g) + NH_3\,(g)$$

	CH$_4$ (g)	N$_2$ (g)	HCN (g)	NH$_3$ (g)
$\Delta_f H^\ominus$/kJ mol^{-1}	−74.81	0.00	135.0	−46.11
$S^\ominus$/J K^{-1} mol^{-1}	186.15	191.5	201.7	192.3

Solution

To calculate the standard reaction Gibbs energy change we use the relationship: $\Delta G^\ominus = \Delta H^\ominus - T\Delta S^\ominus$. This requires finding the enthalpy change in the reaction, $\Delta H^\ominus$, and the entropy change, $\Delta S^\ominus$, from the information given in the table.

The enthalpy change in the reaction is given by:

$$\Delta_r H^\ominus = \Delta_f H^\ominus_{products} - \Delta_f H^\ominus_{reactants}$$

To find the overall enthalpy change for the reaction we insert the enthalpies of formation of the products (HCN and NH$_3$) and subtract those of the reactants (CH$_4$ and N$_2$):

$$\Delta H^\ominus_r = \Sigma \Delta_f H^\ominus(products) - \Sigma \Delta_f H^\ominus(reactants)$$
$$= \Sigma \Delta_f H^\ominus(HCN \text{ and } NH_3) - \Sigma \Delta_f H^\ominus(CH_4 \text{ and } N_2)$$
$$\Delta_r H^\ominus = (135.05 - 46.11) \text{ kJ mol}^{-1} - (-74.81 + 0) \text{ kJ mol}^{-1}$$
$$= (88.89 + 74.81) \text{ kJ mol}^{-1}$$

$$\Delta_r H^\ominus = 163.7 \text{ kJ mol}^{-1}$$

To find the overall entropy change in the reaction we insert the absolute entropy values of the products at 298 K and subtract those of the reactants at 298 K using:

$$\Delta_r S^\ominus = \Sigma v S^\ominus_m(products) - \Sigma v S^\ominus_m(reactants)$$
$$= (201.7 + 192.3) \text{ J K}^{-1} \text{ mol}^{-1} - (186.15 + 191.5) \text{ J K}^{-1} \text{ mol}^{-1}$$
$$= (394 - 377.65) \text{ J K}^{-1} \text{ mol}^{-1}$$
$$= 16.35 \text{ J K}^{-1} \text{ mol}^{-1}$$

We now have the two values required for the Gibbs equation. We are given the temperature so all the parameters can be inserted into the equation:

$$\Delta_r G^\ominus = \Delta_r H^\ominus - T\Delta_r S^\ominus = (163.7 \times 10^3 \text{ J mol}^{-1}) - (298 \text{ K}) \times (16.35 \text{ J K}^{-1} \text{ mol}^{-1}) = 158.83 \times 10^3 \text{ J mol}^{-1}$$
$$= 159 \text{ kJ mol}^{-1}$$

> The standard reaction free energy, $\Delta_r G^\ominus$, is the difference between the molar free energy of the products, $\Sigma G^\ominus_m$ (products) and the molar free energy of the reactants, $\Sigma G^\ominus_m$ (reactants) and is expressed by : $\Delta_r G^\ominus = \Sigma v G^\ominus_m$ (products) $- \Sigma v G^\ominus_m$ (reactants). However as molar free energies are not defined we have to use standard enthalpies of formation and standard reaction enthalpies and apply the Gibbs equation for standard conditions: $\Delta G^\ominus = \Delta H^\ominus - T\Delta S^\ominus$

> The reactant N$_2$ (g) is a pure element and the enthalpy of formation of a pure element in its standard state is zero.

> Because the enthalpy values are in kJ and entropy values in J, the enthalpy value must be multiplied by 10^3 to ensure the units are equivalent.

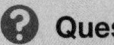

Question 2.27

Formaldehyde (methanal) can be formed according to the following reaction:

H$_2$ (g) + CO (g) → H$_2$CO (g)

$\Delta_r H^\ominus = +1.96$ kJ mol^{-1}

$\Delta_r S^\ominus = -110$ J K^{-1} mol^{-1}

For a reaction temperature of 298 K, calculate the standard Gibbs free energy change, $\Delta_r G^\ominus$, for this reaction and state whether this reaction is spontaneous at this temperature, assuming standard conditions.

Question 2.28

Calculate $\Delta_r G^\ominus$ at 25 °C for the following reaction, using the data below and state whether the reaction will be spontaneous or not at this temperature.

$CH_3COOH\ (l) + 2\ O_2\ (g) \rightarrow 2\ CO_2\ (g) + 2\ H_2O\ (g)$

	CH_3COOH (l)	O_2 (g)	CO_2 (g)	H_2O (g)
$\Delta_f H^\ominus$/kJ mol^{-1}	−484.5	0.00	−393.5	−241.8
$S^\ominus$/J K^{-1} mol^{-1}	159.8	205.1	213.7	188.8

Worked example 2.8I

Predict whether the following reaction is spontaneous at (a) 25 °C and (b) 500 °C:

$N_2\ (g) + 3\ H_2\ (g) \rightleftharpoons 2NH_3\ (g)$

	$\Delta_f H^\ominus$/kJ mol^{-1}	$S^\ominus$/J K^{-1} mol^{-1}
N_2 (g)	0.00	191.6
H_2 (g)	0.00	130.7
NH_3 (g)	−46.11	192.5

Solution

Here we need to use the relationship for Gibbs free energy to determine the sign of this quantity given the values of $\Delta H^\ominus$ and $\Delta S^\ominus$ for the reactants and products. If the sign of $\Delta G^\ominus$ is negative the reaction will be spontaneous.

$\Delta G^\ominus = \Delta H^\ominus - T\Delta S^\ominus$

First calculate $\Delta_r H^\ominus$ and $\Delta S^\ominus$ for the reaction:

$\Delta_r H^\ominus = (2 \times -46.11)\ kJ\ mol^{-1} - (0 + 3 \times 0)\ kJ\ mol^{-1} = -92.22\ kJ\ mol^{-1}$

$\Delta_r S^\ominus = (2 \times 192.5)\ J\ K^{-1}\ mol^{-1} - (191.6 + 3 \times 130.7)\ J\ K^{-1}\ mol^{-1} = -199\ J\ K^{-1}\ mol^{-1}$

a) At 25 °C

For a temperature of 25 °C we must convert to 298 K by adding 273 then insert the $\Delta_r H^\ominus$ and $\Delta S^\ominus$ values into the equation for the Gibbs free energy.

$\Delta_r G^\ominus = -92.22\ kJ\ mol^{-1} - (298\ K \times -0.199\ kJ\ K^{-1}\ mol^{-1}) = -32.9\ kJ\ mol^{-1}$

$\Delta_r G^\ominus$ is negative so the reaction is spontaneous at this temperature.

b) At 500 °C

Again convert the temperature to kelvin by adding 273 to get a temperature of 773 K.

$\Delta_r G^\ominus = -92.22\ kJ\ mol^{-1} - (773\ K \times -0.199\ kJ\ K^{-1}\ mol^{-1}) = +61.6\ kJ\ mol^{-1}$

$\Delta_r G^\ominus$ is positive so the reaction is not spontaneous at this temperature.

In the above calculation we have assumed that $\Delta_f H^\ominus$ and $\Delta_r S^\ominus$ are the same at 298 K and 773 K. This assumption allows us to make a prediction about whether the reaction is spontaneous or not but does not produce accurate values of $\Delta_r G^\ominus$. In order to obtain accurate values of $\Delta_r G^\ominus$ at different temperatures we need to use the **Kirchhoff Equation**:

$\Delta_r H^\ominus_{T2} = \Delta_r H^\ominus_{T1} + \Delta C_p \Delta T$

2.8 ENTROPY AND THE SECOND LAW OF THERMODYNAMICS

$\Delta_r H^\ominus_{T2}$ and $\Delta_r H^\ominus_{T1}$ are the standard enthalpies of reaction at the two temperatures, T_2 and T_1 respectively, and ΔC_p is the difference in heat capacities between the products and reactants. This can be calculated from the equation:

$$\Delta C_p = \Sigma v_i C_p(\text{products}) - \Sigma v_i C_p(\text{reactants})$$

(v_i is the number of moles of each reactant and product in the stoichiometric equation).

> It is usual to assume that ΔC_p does not change with temperature.

Worked example 2.8J

Using the reaction in Worked example 2.8I calculate the actual values for $\Delta_r G^\ominus$ at 500 °C:

$$N_2(g) + 3 H_2(g) \rightleftharpoons 2 NH_3(g)$$

	$\Delta_f H^\ominus_{298}$/kJ mol^{-1}	$S^\ominus_{298}$/J K^{-1} mol^{-1}	C_p/J K^{-1} mol^{-1}
N$_2$ (g)	0	191.6	29.1
H$_2$ (g)	0	130.7	28.8
NH$_3$ (g)	−46.11	192.5	35.7

Solution

Using the equation: $\Delta_r H^\ominus_{T2} = \Delta_r H^\ominus_{T1} + \Delta C_p \Delta T$ to find $\Delta_r H^\ominus$ at the higher temperature of 500 °C we first of all need to calculate ΔC_p.

$\Delta C_p = \Sigma v_i C_p(\text{products}) - \Sigma v_i C_p(\text{reactants})$

$\Delta C_p = (2 \times 35.7 \text{ J K}^{-1} \text{ mol}^{-1}) - [(29.1 \text{ J K}^{-1} \text{ mol}^{-1}) + (3 \times 28.8 \text{ J K}^{-1} \text{ mol}^{-1})]$

$\Delta C_p = 71.4 \text{ J K}^{-1} \text{ mol}^{-1} - 115.5 \text{ J K}^{-1} \text{ mol}^{-1} = -44.1 \text{ J K}^{-1} \text{ mol}^{-1}$

Now convert $\Delta_r H^\ominus$ to the higher temperature using $\Delta_r H^\ominus_{T2} = \Delta_r H^\ominus_{T1} + \Delta C_p \Delta T$

$\Delta H^\ominus_{773} = -92.22 \text{ kJ mol}^{-1} + (-44.1 \text{ J K}^{-1} \text{ mol}^{-1} \times (773 - 298) \text{ K})$

$= -92.22 \text{ kJ mol}^{-1} - 20.95 \text{ kJ mol}^{-1} = -113.2 \text{ kJ mol}^{-1}$

> The temperature of 500 °C must be converted to K by addition of 273 to give us a temperature of 773 K.

$\Delta_r S^\ominus$ must be converted to the appropriate value at the higher temperature by using:

$$\Delta_r S^\ominus = \Delta C_p \ln \frac{T_f}{T_i}$$

So inserting the values of $\Delta_r S^\ominus$ at 500 °C (773 K) we obtain:

$$\Delta_r S^\ominus_{773} = \Delta_r S^\ominus_{298} + \Delta C_p \ln \frac{773 \text{ K}}{298 \text{ K}}$$

Using ΔC_p as calculated above and $\Delta_r S^\ominus$ at 298K from the calculation in Worked example 2.9I:

$\Delta_r S^\ominus_{773} = -198.7 \text{ J K}^{-1} \text{ mol}^{-1} + \left(-44.1 \text{ J K}^{-1} \text{ mol}^{-1} \ln \frac{773 \text{K}}{298 \text{K}}\right)$

$= -240.7 \text{ J K}^{-1} \text{ mol}^{-1}$

> We have seen from the calculations that although $\Delta G^\ominus$ is negative and hence the reaction is spontaneous at room temperature, at higher temperatures $\Delta G^\ominus$ becomes positive and the reaction is no longer spontaneous. The reaction between nitrogen and hydrogen to give ammonia is exothermic and a decrease in temperature favours the forward reaction.

We can now use the Gibbs equation to derive the Gibbs energy at this higher temperature:

$\Delta_r G^\ominus = \Delta_r H^\ominus_{773} - T \Delta S^\ominus_{773}$

$= -113.2 \text{ kJ mol}^{-1} - (773 \text{ K} \times -240.7 \times 10^{-3} \text{ kJ K}^{-1} \text{ mol}^{-1})$

$= +72.9 \text{ kJ mol}^{-1}$

> $\Delta_r S^\ominus$ at 500 °C has been converted from J to kJ.

If this is compared with the approximate value found in 2.8I (b) of +61.4 kJ mol^{-1} we can note the difference when accurate values are used for $\Delta_r H^\ominus$ and $\Delta_r S^\ominus$.

Question 2.29

Calculate $\Delta_r G^{\ominus}$ for the following reaction at 25 °C using values for standard free energies of formation and entropies as given in the table.

$$2\,SO_2(g) + O_2(g) \rightarrow 2\,SO_3(g)$$

	$\Delta H_f^{\ominus}$/kJ mol^{-1}	$S^{\ominus}$/J K^{-1} mol^{-1}
SO$_2$ (g)	−297	248
O$_2$ (g)	0.00	205
SO$_3$ (g)	−395	256

Gibbs free energy and equilibrium

For a reaction to be spontaneous under certain conditions of temperature and pressure the standard Gibbs free energy change in the reaction must be negative, or less than 0, i.e. $\Delta_r G^{\ominus} < 0$.

In reactions that do not go to completion but reach an equilibrium (see Chapter 3) the Gibbs energy decreases until the equilibrium position is reached. So the minimum Gibbs energy for the reaction under the same conditions is found at the point of equilibrium. If the reaction were to continue, the Gibbs energy would increase and the change in Gibbs energy would no longer be negative: the reaction in this direction would not be spontaneous.

At the point at which a reaction reaches equilibrium the Gibbs energy is at a minimum. The concentrations of reactants and products (or partial pressures in a gaseous reaction) at equilibrium are related by the equilibrium constant, K. So there must be a link between the standard Gibbs energy for a reaction and the equilibrium constant.

If $\Delta_r G^{\ominus}$ is a minimum at the point of equilibrium it must be inversely related to the equilibrium constant. The relationship between the Gibbs energy for a reaction and the equilibrium constant is given by:

$$\Delta_r G^{\ominus} = -RT \ln K$$

> To read further about the relationship between Gibbs energy and equilibrium see Chapter 3.

So the larger the value of the equilibrium constant (i.e. the further over to the products side the equilibrium lies) then the more negative (and therefore spontaneous) the reaction. This is sometimes known as the **reaction isotherm** and again its derivation can be found in most standard textbooks.

Worked example 2.8K

Use the following data for Gibbs free energies of formation under standard conditions to calculate the acid-dissociation equilibrium constant (K_a) for formic acid at 25 °C:

Compound	$\Delta_f G^{\ominus}$ (kJ mol^{-1})
HCOOH (aq)	−372.3
H$^+$ (aq)	0.00
HCOO$^-$ (aq)	−351.0

Solution

We must first write the appropriate equation for the equilibrium reaction:

$$HCOOH\,(aq) \rightleftharpoons HCOO^-\,(aq) + H^+\,(aq)$$

$\Delta_r G^{\ominus}$ is given by $\Delta_r G^{\ominus} = \Sigma v \Delta_f G^{\ominus}(\text{products}) - \Sigma v \Delta_f G^{\ominus}(\text{reactants})$

$\Delta_r G^{\ominus} = -351.0\ \text{kJ mol}^{-1} - (-372.3\ \text{kJ mol}^{-1})$

$\qquad = +21.3\ \text{kJ mol}^{-1}$

Now we must rearrange the equation: $\Delta_r G^\ominus = -RT \ln K$ to get an expression for $\ln K$.

$$\ln K = -\frac{\Delta_r G^\ominus}{RT} = -\frac{+21.3 \times 10^3 \text{ J mol}^{-1}}{8.314 \text{ J K}^{-1} \text{ mol}^{-1} \times 298 \text{ K}} = -8.597$$

$$K = 1.85 \times 10^{-4}$$

→ Convert the energy units to joules and the temperature to kelvin.

The small value for K and the positive value for the standard Gibbs free energy suggest that this reaction is not spontaneous.

Question 2.30

The value of $\Delta_r G^\ominus$ for cyclohexane is $+26.8$ kJ mol^{-1}. What is K for this reaction? Write an equation for the formation of cyclohexane from its elements and indicate in which direction the equilibrium lies.

Question 2.31

(a) The water-gas shift reaction provides a way to convert unwanted CO into H_2:

$$H_2O\,(g) + CO\,(g) \rightleftharpoons H_2\,(g) + CO_2\,(g)$$

i. Using the following data for Gibbs free energies of formation, calculate $\Delta_r G^\ominus$ for the reaction at 298 K.

Substance	$\Delta_f G^\ominus$/kJ mol^{-1}
H_2O (g)	−229
CO (g)	−137
H_2 (g)	0.00
CO_2 (g)	−395

ii. Calculate the equilibrium constant, K.

(b) Using the following data for $\Delta_f H^\ominus$ and $S^\ominus$, calculate the 'break even temperature', T, at which $\Delta_r G^\ominus = 0$. State any assumptions you make.

	H_2O (g)	CO (g)	H_2 (g)	CO_2 (g)
$\Delta_f H^\ominus$/kJ mol^{-1}	−242	−111	0	−394
$S^\ominus$/J K^{-1} mol^{-1}	+189	+198	+131	+214

Turn to the Synoptic questions section on page 171 to attempt questions that encourage you to draw on concepts and problem-solving strategies from several topics within a given chapter to come to a final answer.

Final answers to numerical questions appear at the end of the book, and full worked solutions appear on the book's website, where you can also find a set of bonus questions for each chapter. Go to www.oxfordtextbooks.co.uk/orc/chemworkbooks/

3
Chemical equilibrium

When reactants are mixed a reaction ensues until equilibrium is established. The term 'dynamic equilibrium' refers to the fact that once equilibrium has been established, both forward and backward reactions continue at the same rate. Reactants and products will both be present but overall there is no further tendency for the mixture to undergo net change. For example, consider the reaction between hydrogen and iodine to produce hydrogen iodide:

$$H_2(g) + I_2(g) \rightleftharpoons 2\,HI(g)$$

The equilibrium process may be investigated by setting up a reaction mixture of hydrogen and iodine in a sealed vessel and allowing it to come to equilibrium or by enclosing the product hydrogen iodide in a sealed vessel and allowing it to come to equilibrium. Both reaction vessels will establish equilibrium whereupon each vessel will contain a fixed ratio of reactants and products.

The ratio of reactants to products found at equilibrium is determined by the equilibrium constant for the reaction. If the ratio is expressed in terms of the concentration of the reactants and products (measured in mol dm^{-3}) then the equilibrium constant, K_c, for the reaction can be expressed as:

$$K_c = \frac{[HI]^2}{[H_2][I_2]}$$

Background reading to cover the theory behind the worked examples is highly recommended.

3.1 Equilibrium and the Gibbs energy minimum

The course of a reaction can be followed in terms of the extent of the reaction, ξ. The extent of the reaction has units of moles and is an indication of its progress: at the beginning of the reaction it has a value of zero. When equilibrium is achieved, the Gibbs energy of the mixture, G, measured in joules, has reached a minimum as shown in Figure 3.1.

$\Delta_r G = \dfrac{dG}{d\xi} = 0$

Forward reaction spontaneous if $\Delta_r G < 0$
Reverse reaction spontaneous if $\Delta_r G > 0$
Reaction is at equilibrium if $\Delta_r G = 0$

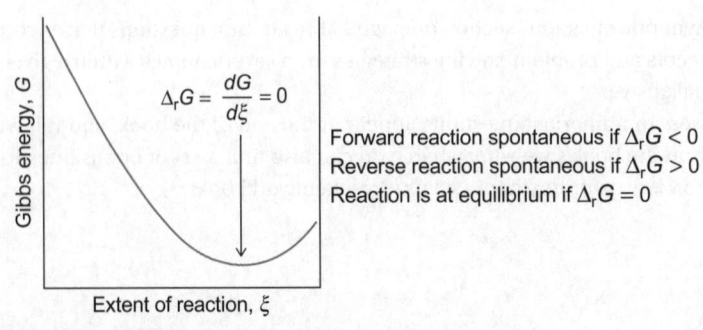

Figure 3.1 Plot of Gibbs energy, G, against the extent of the reaction, ξ. The reaction Gibbs energy, $\Delta_r G$, is the gradient of this plot. The position of equilibrium corresponds to the minimum in this plot where the gradient is zero.

On the plot of Gibbs energy, G, versus extent of reaction, ξ, such as that shown in Figure 3.1, the position of equilibrium corresponds to the point at which the gradient is zero. $\Delta_r G$ is the derivative of the Gibbs energy of the reaction, G, with respect to the extent of the reaction, ξ, and at a given temperature, T, and pressure, p:

$$\Delta_r G = \left(\frac{dG}{d\xi} \right)_{p,T}$$

hence, $\Delta_r G$ is defined as the gradient of the graph of the Gibbs energy, G, plotted against the extent of the reaction, ξ. The forward reaction is spontaneous when $\Delta_r G$ is less than zero, and the reverse reaction is spontaneous if $\Delta_r G$ is greater than zero. The reaction is at equilibrium if $\Delta_r G$ equals zero. Remember, of course, that there is no correlation between the degree of spontaneity of a reaction and its rate.

The reaction Gibbs energy, $\Delta_r G$, is the difference between the chemical potentials of the reactants and products. As the chemical potentials vary with composition, $\Delta_r G$ can vary as the reaction proceeds. The reaction Gibbs energy, $\Delta_r G$, should not be confused with the standard Gibbs energy of reaction, $\Delta_r G^\ominus$, which is defined as the difference in standard molar Gibbs energies of the reagents and products. These two quantities are related by the following expression:

$$\Delta_r G = \Delta_r G^\ominus + RT \ln Q$$

where $\Delta_r G$ is the reaction Gibbs energy measured in J mol^{-1}, $\Delta_r G^\ominus$ is the standard Gibbs energy of reaction measured in J mol^{-1}, R is the gas constant measured in J K^{-1} mol^{-1}, T is the temperature measured in K, and Q is the reaction quotient (a dimensionless quantity).

At equilibrium $\Delta_r G$ is zero and the reaction quotient is referred to as the equilibrium constant, yielding the following expression:

$$\Delta_r G = \Delta_r G^\ominus + RT \ln Q$$

$$0 = \Delta_r G^\ominus + RT \ln K$$

$$\Delta_r G^\ominus = -RT \ln K$$

where $\Delta_r G^\ominus$ is the standard Gibbs energy of reaction measured in J mol^{-1}, R is the gas constant measured in J K^{-1} mol^{-1}, T is the temperature measured in K, and K is the equilibrium constant (a dimensionless quantity).

The following section covers many different types of equilibria including: gaseous reactions; acids and bases; sparingly soluble salts; and complex ion formation. Equilibria in which the reactants and products are all in the same phase are known as homogeneous equilibria, an example being a gaseous system. Equilibria in which the reactants and products are in a mixture of different phases are known as heterogeneous equilibria, an example being the dissolution of a sparingly soluble salt.

The position of equilibrium may be quantified by determining a value for the equilibrium constant for the reaction. Generally, the larger the value of the equilibrium constant, the greater is the relative concentration of products to reactants in the equilibrium mixture. Similarly, the smaller the value of the equilibrium constant, the lesser is the relative concentration of products to reactants in the equilibrium mixture. The exact ratio of reactants to products is determined by the expression for the equilibrium constant. Remember that a discussion of equilibria does not involve any reference to the rate of a reaction; it is entirely possible for a spontaneous reaction to occur at a slow rate and thus take a long time to achieve equilibrium.

3.2 The equilibrium constant—approximations for real systems

For a reaction of the general form:

$$x\,A + y\,B \rightleftharpoons m\,C + n\,D$$

→ $\Delta_r G$ is defined as the slope of the graph of the Gibbs energy, G, plotted against the extent of the reaction, ξ. Take care not to confuse the reaction Gibbs energy, $\Delta_r G$, with the standard Gibbs energy of reaction, $\Delta_r G^\ominus$.

→ Chemical potential represents the potential that a substance has for undergoing physical or chemical change. For a one-component system the chemical potential is the molar Gibbs energy, for a mixture the chemical potential is the partial molar Gibbs energy. A system at equilibrium has uniform chemical potential.

→ ln Q represents the natural logarithm of the reaction quotient, the ratio of activities of the products to the reagents. The reaction quotient is dimensionless.

→ When the symbol $\ominus$ is associated with a thermodynamic parameter it refers to the standard state (1 bar pressure). Note that temperature is not part of the definition of standard state.

→ Note that an equation for the reaction must always be provided when discussing equilibrium.

> The activity of a species is its effective concentration. Note that activity is a dimensionless quantity.

the thermodynamic equilibrium constant K may be expressed in terms of the activities, a, of the reactants and products at equilibrium:

$$K = \frac{(a_C)^m (a_D)^n}{(a_A)^x (a_B)^y}$$

For a perfect gas, the activity, a, is the ratio of its partial pressure, p (measured in bar), to the standard state pressure, $p^\ominus$ (measured in bar); for perfect solutes in solution the activity, a, is the ratio of the molar concentration (measured in mol dm^{-3}) to the standard concentration (measured in mol dm^{-3}). So:

$$K = \frac{(a_C)^m (a_D)^n}{(a_A)^x (a_B)^y} = \frac{\left(\frac{p_C}{p^\ominus}\right)^m \left(\frac{p_D}{p^\ominus}\right)^n}{\left(\frac{p_A}{p^\ominus}\right)^x \left(\frac{p_B}{p^\ominus}\right)^y}$$

or:

$$K = \frac{(a_C)^m (a_D)^n}{(a_A)^x (a_B)^y} = \frac{\left(\frac{[C]}{[C]^\ominus}\right)^m \left(\frac{[D]}{[D]^\ominus}\right)^n}{\left(\frac{[A]}{[A]^\ominus}\right)^x \left(\frac{[B]}{[B]^\ominus}\right)^y}$$

Equilibrium constant expressions for heterogeneous systems are simplified because the concentrations (or activities) of pure solids and pure liquids are constant at a given temperature, and are equal to 1. Also note that K is a unitless quantity.

However, it is common practice to make approximations to these expressions. For a gaseous system at equilibrium we consider the ratio of the partial pressures of the products and reactants and refer to the constant as K_p:

$$K_p = \frac{(p_C)^m (p_D)^n}{(p_A)^x (p_B)^y}$$

and for solutes in solution at equilibrium we consider the ratio of the concentrations of the products and reagents and refer to the constant as K_c:

$$K_c = \frac{[C]^m [D]^n}{[A]^x [B]^y}$$

K_p and K_c are useful approximations and work well for dilute mixtures. Their units depend upon the stoichiometry of the reaction.

Table 3.1 summarizes the different types of equilibrium constant.

Table 3.1 Summary of the different types of equilibrium constant.

K	K_p	K_c
Dimensionless	Units of partial pressure (exact units depend upon the expression for K_p and p, and sometimes all terms cancel out)	Units of concentration (exact units depend upon the expression for K_c and sometimes all terms cancel out)
Thermodynamic equilibrium constant expressed in terms of activities	Equilibrium constant expressed in terms of partial pressures (a valid approximation when the partial pressures are low and the gases behave perfectly)	Equilibrium constant expressed in terms of concentration (a valid approximation when the concentrations are low and the solutions behave ideally)

Worked example 3.2A

Consider the equilibrium between nitrogen and hydrogen to form ammonia:

$$N_2(g) + 3H_2(g) \rightleftharpoons 2NH_3(g)$$

Give expressions for K, K_p, and K_c and include their units.

Solution

The reactants (nitrogen and hydrogen) are on the left-hand side of the equation and the product (ammonia) is on the right-hand side. The number of atoms of each element on the reactant side balances with the number of atoms of each element on the product side so the equation is balanced. Expressions for K, K_p, and K_c may now be given:

$$K = \frac{(a_{NH_3})^2}{(a_{N_2})(a_{H_2})^3} \text{ with no units}$$

$$K_p = \frac{(p_{NH_3})^2}{(p_{N_2})(p_{H_2})^3} \text{ with units of atm}^{-2}$$

→ Note that other units for pressure may be used.

and

$$K_c = \frac{[NH_3]^2}{[N_2][H_3]^3} \text{ with units of mol}^{-2} \text{ dm}^6$$

→ Check that products are placed as the numerator and reagents are placed as the denominator. Also pay particular attention to the stoichiometry of the reaction and use the correct exponents. In this case the ammonia term is raised to the power 2 and the hydrogen term is raised to the power 3 according to the stoichiometric coefficients.

Worked example 3.2B

Consider the equilibrium between nickel and carbon monoxide to form nickel carbonyl. At a temperature greater than the melting point of nickel carbonyl (43 °C) the product will be gaseous:

$$Ni(s) + 4CO(g) \rightleftharpoons Ni(CO)_4(g)$$

Give expressions for K, K_p, and K_c and state the units where appropriate.

Solution

The reactants (nickel and carbon monoxide) are on the left-hand side of the equation and the product (nickel carbonyl) is on the right-hand side. The number of atoms of each element on the reactant side balances with the number of atoms of each element on the product side so the equation is balanced. It is noted that nickel is present in solid form and hence will not form part of the equilibrium constant expressions as its activity is equal to 1. Expressions for K, K_p, and K_c may now be given:

$$K = \frac{(a_{Ni(CO)_4})}{(a_{Ni})(a_{CO})^4} = \frac{(a_{Ni(CO)_4})}{(a_{CO})^4} \text{ with no units}$$

$$K_p = \frac{p_{Ni(CO)_4}}{(p_{CO})^4} \text{ with units of atm}^{-3}$$

→ Note that other units for pressure may be used.

and

$$K_c = \frac{[Ni(CO)_4]}{[CO]^4} \text{ with units of mol}^{-3} \text{ dm}^9$$

> **Question 3.1**
>
> Give expressions for K, K_p, and K_c for the following reactions and state the units where appropriate.
>
> (a) $H_2(g) + I_2(g) \rightleftharpoons 2HI(g)$
>
> (b) $2SO_3(g) \rightleftharpoons 2SO_2(g) + O_2(g)$
>
> (c) $PCl_5(g) \rightleftharpoons PCl_3(g) + Cl_2(g)$
>
> (d) $2H_2S(g) \rightleftharpoons 2H_2(g) + S_2(s)$
>
> (e) $N_2(g) + O_2(g) \rightleftharpoons 2NO(g)$

3.3 How K_p, K_c, and K are related

K_p and K_c are related to each other according to the following expression:

$$K_p = K_c(RT)^{\Delta n}$$

where:

- K_p is the equilibrium constant based on partial pressures.
- K_c is the equilibrium constant based on concentrations.
- R is the gas constant (various values and units can be used).
- T is the temperature (K).
- Δn is the change in amount in moles of gaseous species during the reaction (i.e. amount in moles of gaseous products – amount in moles of gaseous reactants).

→ The units of K_p will depend upon the value of R used.

If $\Delta n = 0$ then K_p and K_c have the same numerical value; if K_p is measured in atm then it has the same numerical value as K.

Worked example 3.3A

Determine K_p (first with units of atm and then with units of Pa) for the equilibrium dissociation of N_2O_4 at 350 K according to the following equation:

$$N_2O_4(g) \rightleftharpoons 2NO_2(g) \quad K_c = 0.13 \text{ mol dm}^{-3}$$

Solution

In this case the amount in moles of gaseous products is 2 and the amount in moles of gaseous reactants is 1. Hence $\Delta n = 1$. So if K_c has a value of 0.13 mol dm^{-3} then K_p can be given with units of atm by:

$$K_p = K_c(RT)^{\Delta n} = (0.13 \text{ mol dm}^{-3}) \times ((0.0821 \text{ dm}^3 \text{ atm K}^{-1} \text{ mol}^{-1}) \times (350 \text{ K}))^1 = 3.7 \text{ atm}$$

or with units of Pa by:

$$K_p = K_c(RT)^{\Delta n} = (130 \text{ mol m}^{-3}) \times ((8.314 \text{ m}^3 \text{ J K}^{-1} \text{ mol}^{-1}) \times (350 \text{ K}))^1 = 370 \text{ kPa}$$

As 1 atm = 1.013×10^5 Pa it can be seen that the two answers are equivalent.

→ Note that K_c must have units of mol m^{-3} in order to arrive at an answer with units of Pa. 1 mol dm^{-3} is 1000 mol m^{-3} so 0.13 mol dm^{-3} = 130 mol m^{-3}. Remember also that 1 J = kg m^2 s^{-2} and 1 Pa = kg m^{-1} s^{-2} and hence the answer has units of Pa.

Worked example 3.3B

Determine K_c for the equilibrium between nitrogen and hydrogen to form ammonia at 25 °C.

$$N_2(g) + 3H_2(g) \rightleftharpoons 2NH_3(g) \quad K_p = 6.10 \times 10^5 \text{ atm}^{-2}$$

Solution

In this case the amount in moles of gaseous products is 2 and the number in moles of gaseous reactants is 4. Hence $\Delta n = -2$. So if K_p has a value of 6.10×10^5 atm^{-2} then K_c can be given by:

$$K_p = K_c (RT)^{\Delta n}$$

$$K_c = \frac{K_p}{(RT)^{\Delta n}} = \frac{(6.10 \times 10^5 \text{ atm}^{-2})}{\left((0.0821 \text{ dm}^3 \text{ atm K}^{-1} \text{ mol}^{-1}) \times (298 \text{ K})\right)^{-2}} = 3.6 \times 10^8 \text{ mol}^{-2} \text{ dm}^6$$

→ Note that the temperature must be converted from degree Celsius to kelvin.

❓ Question 3.2

For the following reactions determine K_c.

(a) $N_2O_4(g) \rightleftharpoons 2NO_2(g)$ $\quad K_p = 1.78 \times 10^4$ atm at 327 °C

(b) $N_2(g) + 3H_2(g) \rightleftharpoons 2NH_3(g)$ $\quad K_p = 40.7$ atm^{-2} at 400 K

(c) $2SO_2(g) + O_2(g) \rightleftharpoons 2SO_3(g)$ $\quad K_p = 4.0 \times 10^{24}$ atm^{-1} at 298 K

(d) $H_2(g) + I_2(g) \rightleftharpoons 2HI(g)$ $\quad K_p = 54$ at 427 °C

3.4 Forward and reverse reactions

Remember that the equation for the reaction under consideration must always be given as the equilibrium constants for the forward and backward reactions are not identical. For the reaction of general form:

$$2A + B \rightleftharpoons 2C + 2D$$

The equilibrium constant denoted K_{c1} may be expressed as:

$$K_{c1} = \frac{[C]^2 [D]^2}{[A]^2 [B]}$$

By contrast, if the reaction is written in reverse in the general form:

$$2C + 2D \rightleftharpoons 2A + B$$

Then the equilibrium constant, now denoted as K_{c2}, may be expressed as:

$$K_{c2} = \frac{[A]^2 [B]}{[C]^2 [D]^2}$$

Comparing the expressions for K_{c1} and K_{c2} it can be seen that they are not equal. They are, however, related:

$$K_{c1} = \frac{1}{K_{c2}}$$

Worked example 3.4A

Consider the following equilibrium where dinitrogen tetroxide dissociates to give nitrogen dioxide:

$$N_2O_4(g) \rightleftharpoons 2\,NO_2(g) \quad K_p = 1.78 \times 10^4 \text{ atm at 600 K}$$

What is the equilibrium constant for the reverse reaction?

Solution

The equilibrium constant for the reverse reaction:

$$2\,NO_2(g) \rightleftharpoons N_2O_4(g)$$

is given by:

$$K_{p\,\text{reverse}} = \frac{1}{K_{p\,\text{forward}}}$$

$$K_{p\,\text{reverse}} = \frac{1}{(1.78 \times 10^4 \text{ atm})}$$

$$K_{p\,\text{reverse}} = 5.62 \times 10^{-5} \text{ atm}^{-1}$$

➔ Note temperature does not form part of the calculation but it is often quoted as part of the reaction description.

❓ Question 3.3

Given the following equilibria, calculate the values of the equilibrium constants for the reverse reactions:

(a) $N_2(g) + 3H_2(g) \rightleftharpoons 2NH_3(g)$ $K_p = 40.7 \text{ atm}^{-2}$ at 400 K

(b) $2SO_2(g) + O_2(g) \rightleftharpoons 2SO_3(g)$ $K_p = 4.0 \times 10^{24} \text{ atm}^{-1}$ at 298 K

(c) $H_2(g) + I_2(g) \rightleftharpoons 2HI(g)$ $K_p = 54$ at 700 K

3.5 Le Chatelier's principle

Le Chatelier's principle states that when a system at equilibrium is disturbed it responds to the disturbance by behaving in a manner that tends to oppose the change.

Changing temperature: At a given temperature K is fixed, but if the reaction temperature is changed then the value of the equilibrium constant changes accordingly: it can increase or decrease depending on whether the reaction is exothermic or endothermic. If temperature is raised then the endothermic reaction is favoured and the composition of the equilibrium mixture changes. If temperature is lowered then the exothermic reaction is favoured and the composition of the equilibrium mixture once again changes.

Changing the pressure by addition of an inert gas: The equilibrium constant is independent of pressure at constant temperature. The partial pressures of the reactants and products are unchanged and there is no effect upon the composition of the equilibrium mixture.

Changing the pressure by altering the volume of the system: The equilibrium constant is independent of pressure at constant temperature. However, if the reaction pressure is changed, either by confining the system to a smaller volume or by expanding the system to a larger volume, then the partial pressures of the reactants and products are changed. For some reactions, the position of equilibrium will change to compensate for this in order to maintain a constant value for the equilibrium constant. For example, if there is a change in number of gaseous molecules during the course of the reaction and pressure is increased, the reaction producing the smaller number of gaseous molecules is favoured. For reactions where there is no change in the number of gaseous molecules then altering pressure has no effect upon the position of equilibrium.

3.6 Standard Gibbs energy change and the position of equilibria

The thermodynamic equilibrium constant for a reaction, K, is related to the standard Gibbs energy change, $\Delta_r G^\ominus$, according to the following expression:

$$\Delta_r G^\ominus = -RT \ln K$$

where:

- $\Delta_r G^\ominus$ is the standard Gibbs energy change (in J mol^{-1}).
- R is the gas constant (8.314 J K^{-1} mol^{-1}).
- T is the temperature (in K).
- K is the thermodynamic equilibrium constant (a unitless number).

$\Delta_r G^\ominus$ is the change in Gibbs energy per mol of a compound formed at 1 bar from its elements in their standard states; it can be a positive or a negative value. Using tables of thermodynamic data this equation can be used to predict the thermodynamic equilibrium constant for a reaction at any temperature but the equation must be rearranged to make $\ln K$ the subject:

$$\ln K = -\frac{\Delta_r G^\ominus}{RT}$$

In order to determine K we need to remove the natural logarithm term from the left-hand side of the expression; the rule used is that if $\ln x = y$ then $x = e^y$.

$$K = e^{-\left(\frac{\Delta_r G^\ominus}{RT}\right)}$$

Values for K can be small or large depending on the sign of $\Delta_r G^\ominus$; small numbers result if $\Delta_r G^\ominus$ is positive and large numbers result if $\Delta_r G^\ominus$ is negative.

→ Note that e is a mathematical constant: it is the base of the natural logarithm and has a value of 2.718. On a scientific calculator you will find an antilog, e^x, function which allows the value of the exponential function to be determined.

Worked example 3.6A

For a given reaction at 25 °C, $\Delta_r G^\ominus = -2.50$ kJ mol^{-1}; calculate the thermodynamic equilibrium constant, K.

Solution

Write the expression for K, replace any known terms and solve to find the value of K.

$$K = e^{-\left(\frac{\Delta_r G^\ominus}{RT}\right)}$$

$$K = e^{-\left(\frac{-2.50 \times 10^3 \text{ J mol}^{-1}}{8.314 \text{ J K}^{-1} \text{ mol}^{-1} \times 298 \text{ K}}\right)} = 2.74$$

→ Remember to convert from degree Celsius to kelvin and remember that $\Delta_r G^\ominus$ should have units of J mol^{-1}. Does the answer make sense? Yes: the standard Gibbs energy change is negative so we expect the answer to be larger than 1.

Worked example 3.6B

For a given reaction at 25 °C, $K = 6.0 \times 10^5$; calculate the standard Gibbs energy change.

> Does the answer make sense? Yes: the equilibrium constant is a large number indicating a spontaneous reaction so the standard Gibbs energy change should be negative.

Solution

Write the expression for $\Delta_r G^\circ$, replace any known terms and solve to find the value of $\Delta_r G^\circ$.

$$\Delta_r G^\circ = -RT \ln K$$

$$\Delta_r G^\circ = -(8.314 \text{ J K}^{-1} \text{ mol}^{-1}) \times (298 \text{ K}) \times (\ln 6.0 \times 10^5)$$

$$\Delta_r G^\circ = -33 \text{ kJ mol}^{-1}$$

> **Question 3.4**
>
> (a) For a given reaction at 25 °C, $\Delta_r G^\circ = 3.250$ kJ mol^{-1}; calculate the thermodynamic equilibrium constant, K.
>
> (b) For a given reaction at 25 °C, $K = 2 \times 10^{-5}$; calculate the standard Gibbs energy change, $\Delta_r G^\circ$.

3.7 Temperature effect

For a given reaction the value of its equilibrium constant varies with temperature; whether it increases or decreases as temperature goes up depends on whether the reaction is endothermic or exothermic. For an endothermic reaction, the value of the equilibrium constant **increases** as temperature increases. For an exothermic reaction, the value of the equilibrium constant **decreases** as temperature increases.

Using the following two expressions the relationship between temperature and the enthalpy change of the reaction can be found:

$$\Delta_r G^\circ = -RT \ln K$$

where:

- $\Delta_r G^\circ$ is the standard Gibbs energy change (in J mol^{-1}).
- R is the gas constant (8.314 J K^{-1} mol^{-1}).
- T is the temperature (in kelvin).
- K is the thermodynamic equilibrium constant (a unitless number).

and:

$$\Delta_r G^\circ = \Delta_r H^\circ - T\Delta_r S^\circ$$

where:

- $\Delta_r G^\circ$ is the standard Gibbs energy change (in J mol^{-1}).
- $\Delta_r H^\circ$ is the standard enthalpy change (in J mol^{-1}).
- T is the temperature (in kelvin).
- $\Delta_r S^\circ$ is the standard entropy change (in J K^{-1} mol^{-1}).

First equate these two expressions:

$$-RT \ln K = \Delta_r H^\circ - T\Delta_r S^\circ$$

and divide through by $-RT$:

$$\ln K = \frac{\Delta_r H^\circ}{-RT} - \frac{T\Delta_r S^\circ}{-RT}$$

which simplifies to:

$$\ln K = -\frac{\Delta_r H^\ominus}{RT} + \frac{\Delta_r S^\ominus}{R}$$

This relationship is known as the van't Hoff equation. Measuring the variation of the equilibrium constant with temperature thus provides a means of determining the standard enthalpy and entropy change for a reaction.

$$\ln K = -\frac{\Delta_r H^\ominus}{RT} + \frac{\Delta_r S^\ominus}{R}$$

This can be rewritten as:

$$\ln K = -\frac{\Delta_r H^\ominus}{R}\frac{1}{T} + \frac{\Delta_r S^\ominus}{R}$$

and if $\ln K$ is plotted against $1/T$ then a linear graph results with a gradient of $-\Delta_r H^\ominus/R$ and an intercept of $\Delta_r S^\ominus/R$, as shown in Figure 3.2.

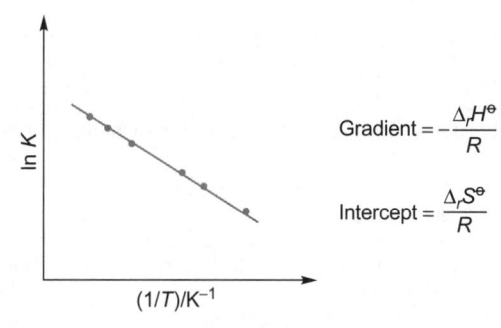

Figure 3.2 A plot of $\ln K$ against $1/T$. The standard enthalpy change can be found from the gradient.

For a given reaction, if we know the standard enthalpy change of reaction and also know the equilibrium constant at one temperature, we can predict a new value for the equilibrium constant at another temperature using the following expression:

$$\ln \frac{K_2}{K_1} = \frac{-\Delta_r H^\ominus}{R}\left(\frac{1}{T_2} - \frac{1}{T_1}\right)$$

where K_1 and K_2 denote the thermodynamic equilibrium constants at temperatures T_1 and T_2 respectively, $\Delta_r H^\ominus$ denotes the standard enthalpy change of the reaction, and R denotes the gas constant (8.314 J K^{-1} mol^{-1}).

→ Note that the standard enthalpy and entropy change for a reaction are assumed to be constant across the temperature change in order to derive this expression.

Worked example 3.7A

The equilibrium between nitrogen and hydrogen to form ammonia at 25 °C is as follows:

$$N_2(g) + 3H_2(g) \rightleftharpoons 2NH_3(g) \quad K = 6.10 \times 10^5$$

The standard enthalpy change, $\Delta_r H^\ominus$, is −92.2 kJ mol^{-1}. Estimate the value of the equilibrium constant at 67 °C.

Solution

Using the expression below, assume that the standard enthalpy change is constant over the range of temperatures and replace any known terms.

$$\ln \frac{K_2}{K_1} = \frac{-\Delta_r H^\ominus}{R}\left(\frac{1}{T_2} - \frac{1}{T_1}\right)$$

3 CHEMICAL EQUILIBRIUM

> Remember to convert from Celsius to kelvin and remember that $\Delta_r H^\ominus$ should have units of J mol^{-1}, hence the units cancel.

$$\ln \frac{K_2}{K_1} = \frac{-(-92.2 \times 10^3 \text{ J mol}^{-1})}{8.314 \text{ J K}^{-1} \text{mol}^{-1}} \left(\frac{1}{340 \text{ K}} - \frac{1}{298 \text{ K}} \right)$$

$$\ln \frac{K_2}{K_1} = -4.60$$

$\ln K_2/K_1$ represents the natural logarithm of K_2/K_1 and to evaluate K_2 we first take the antilogarithm:

> If $\ln K_2 / K_1 = -x$ then $K_2 / K_1 = e^{-x}$

> Does the answer make sense? Yes: the standard enthalpy change is negative so the reaction is exothermic and we expect the equilibrium constant to get smaller as temperature is increased.

$$\frac{K_2}{K_1} = e^{-4.60} = 0.010$$

and then rearrange the expression to find K_2:

$$K_2 = (0.010 \times 6.10 \times 10^5) = 6.1 \times 10^3$$

$$K_2 = 6.1 \times 10^3$$

Question 3.5

(a) The equilibrium between nitrogen and hydrogen to form ammonia at 25 °C is as follows:

$$N_2(g) + 3H_2(g) \rightleftharpoons 2NH_3(g) \quad K = 6.1 \times 10^5$$

The standard enthalpy change, $\Delta_r H^\ominus$, is -92.2 kJ mol^{-1}. Estimate the value of the equilibrium constant at 350 °C.

(b) The equilibrium between sulfur dioxide and oxygen to form sulfur trioxide at 25 °C is as follows:

$$2SO_2(g) + O_2(g) \rightleftharpoons 2SO_3(g) \quad K = 4.0 \times 10^{24}$$

The standard enthalpy change, $\Delta_r H^\ominus$, is -198 kJ mol^{-1}. Estimate the value of the equilibrium constant at 700 °C.

3.8 Calculating equilibrium constants and equilibrium compositions

If the equilibrium concentrations, or equilibrium partial pressures, for a given reaction are known then it is possible to determine a value for K_c or K_p. Similarly values of K_c and K_p can be used to determine the equilibrium concentrations or equilibrium partial pressures for a reaction. However, a good understanding of algebra is required in order to solve most problems; the following examples attempt to demonstrate the breadth of this area.

Worked example 3.8A

Consider the equilibrium between hydrogen and iodine to form hydrogen iodide:

$$H_2(g) + I_2(g) \rightleftharpoons 2HI(g)$$

Pure hydrogen and pure iodine are mixed together at 700 K and are allowed to reach equilibrium. The equilibrium concentrations of hydrogen and iodine are each found to be 4.9 mol dm^{-3}; if K_c has a value of 54, what is the equilibrium concentration of hydrogen iodide?

Solution

Begin by writing an expression for K_c:

$$K_c = \frac{[HI]^2}{[H_2][I_2]}$$

Next, consider the information provided in the question and replace any known terms:

$$K_c = \frac{[HI]^2}{[H_2][I_2]} = \frac{[HI]^2}{(4.9 \text{ mol dm}^{-3}) \times (4.9 \text{ mol dm}^{-3})} = 54$$

Rearrange the expression to make $[HI]^2$ the subject:

$$[HI]^2 = 54 \times (4.9 \text{ mol dm}^{-3}) \times (4.9 \text{ mol dm}^{-3})$$

Then take the square root of both sides to determine $[HI]$:

$$[HI] = \sqrt{54 \times (4.9 \text{ mol dm}^{-3}) \times (4.9 \text{ mol dm}^{-3})} = 36 \text{ mol dm}^{-3}$$

The equilibrium concentration of hydrogen iodide is 36 mol dm^{-3}.

→ Check that products are placed as the numerator and reagents are placed as the denominator. Also pay particular attention to the stoichiometry of the reaction and use the correct exponents. There are no units associated with the equilibrium constant because the stoichiometry of the reaction ensures that the concentration terms in the numerator and denominator cancel.

→ Note that the value for the equilibrium constant and the equilibrium concentrations of the reagents are given in the question.

→ Check that the answer makes sense by inserting the values for all equilibrium concentrations back into the expression for K_c, check that a value of 54 is obtained; this is the case so the answer is sensible.

Worked example 3.8B

Consider the equilibrium between hydrogen and iodine to form hydrogen iodide:

$$H_2(g) + I_2(g) \rightleftharpoons 2HI(g)$$

If 4.00 mol of hydrogen are reacted with 4.00 mol of iodine in a 2.00 dm^3 vessel and K_c has a value of 54.0 what is the composition of the equilibrium mixture?

→ Note many different combinations of moles and volumes are possible in a question such as this, sometimes the concentration terms are directly provided.

Solution

In order to answer the question we have to determine the value of $[HI]$, $[H_2]$, and $[I_2]$ at equilibrium. However, the question only supplies information regarding the starting amount in moles of reactants, the volume of the reaction vessel and the value of K_c, hence multiple steps are required in order to get to the answer. Here the answer is broken down into five steps.

Step 1: Begin by writing the expression for K_c, consider the information provided in the question, and replace any known terms:

$$K_c = \frac{[HI]^2}{[H_2][I_2]} = 54.0$$

Step 2: Information about the initial amounts of the reagents and the volume of the reaction vessel is provided in the question. Use this to determine the initial concentration of the two reagents:

$$[H_2]_{initial} = \frac{4.00 \text{ mol}}{2.00 \text{ dm}^3} = 2.00 \text{ mol dm}^{-3}$$

$$[I_2]_{initial} = \frac{4.00 \text{ mol}}{2.00 \text{ dm}^3} = 2.00 \text{ mol dm}^{-3}$$

Step 3: Set up an equilibrium table.

	H$_2$	I$_2$	HI
Initial concentration	2.00 mol dm^{-3}	2.00 mol dm^{-3}	0.00 mol dm^{-3}
Change in concentration	x mol dm^{-3}	$-x$ mol dm^{-3}	$+2x$ mol dm^{-3}
Equilibrium concentration	$(2.00 - x)$ mol dm^{-3}	$(2.00 - x)$ mol dm^{-3}	$(0.00 + 2x)$ mol dm^{-3}

→ It is recommended that the following style of equilibrium table is used in order to keep track of moles reacted and formed, and let x represent the number of moles dm^{-3} of reagent that has reacted at equilibrium.

Initial concentration is the concentration present at the start of the reaction.

The change in concentration is determined by the stoichiometry of the reaction. In this case 1 mol of hydrogen reacts with 1 mol of iodine to form 2 mol of hydrogen iodide. So $2x$ mol of hydrogen iodide are formed at equilibrium.

The equilibrium concentrations are given by the sum of the initial concentration term and the change in concentration term for each reactant and product.

Step 4: In order to answer the question we must determine the value of x in order to solve the bottom row of the table. Take the expression for K_c, and replace the terms on the right-hand side with the expressions from the bottom row of the equilibrium table and solve for x:

$$K_c = \frac{[HI]^2}{[H_2][I_2]}$$

$$54.0 = \frac{(0.00+2x)^2}{(2.00-x)(2.00-x)}$$

This expression can be simplified to give:

$$54.0 = \frac{(2x)^2}{(2.00-x)^2}$$

and now solve for x:

$$\sqrt{54.0} = \frac{\sqrt{(2x)^2}}{\sqrt{(2.00-x)^2}}$$

taking the sign of the root

$$7.35 = \frac{2x}{2.00-x}$$

$$7.35(2.00-x) = 2x$$

$$14.70 - 7.35x = 2x$$

$$14.70 = 2x + 7.35x$$

$$14.70 = 9.35x$$

$$\frac{14.70}{9.35} = x$$

$$x = 1.57$$

→ Remember that x represents the amount in mol dm^{-3} of reagent that has reacted at equilibrium.

→ Check that the answer makes sense by inserting the equilibrium concentrations back into the expression for K_c. Doing so yields a value of 53 for K_c, very close to the value quoted in the question, so the answer is sensible.

Step 5: Determine the composition of the equilibrium mixture:

at equilibrium $[H_2] = (2.00 - x)$ mol dm^{-3} = $(2.00 - 1.57)$ mol dm^{-3} = 0.43 mol dm^{-3};

at equilibrium $[I_2] = (2.00 - x)$ mol dm^{-3} = $(2.00 - 1.57)$ mol dm^{-3} = 0.43 mol dm^{-3};

at equilibrium $[HI] = (0.00 + 2x)$ mol dm^{-3} = $(0.00 + 2(1.57))$ mol dm^{-3} = 3.14 mol dm^{-3}.

→ At first glance this question may seem similar to the previous example. However, the mathematical processes used to determine the answer are different.

Worked example 3.8C

Consider the equilibrium between hydrogen and iodine to form hydrogen iodide:

$$H_2(g) + I_2(g) \rightleftharpoons 2HI(g)$$

If 4.00 mol of hydrogen are reacted with 6.00 mol of iodine in a 2.00 dm^3 vessel and K_c has a value of 54 what is the composition of the equilibrium mixture?

Solution

In order to answer the question once again we have to determine the value of [HI], [H$_2$], and [I$_2$] at equilibrium. However, the question only supplies information regarding the starting amount in moles of reagents, the volume of the reaction vessel, and the value of K_c, hence multiple steps are required in order to get to the answer. Here the answer is broken down into five steps.

Step 1: Begin by writing the expression for K_c, consider the information provided in the question and replace any known terms.

$$K_c = \frac{[HI]^2}{[H_2][I_2]} = 54$$

3.8 CALCULATING EQUILIBRIUM CONSTANTS AND EQUILIBRIUM COMPOSITIONS

Step 2: Information about the initial amounts of the reagents and the volume of the reaction vessel is provided in the question. Use this to determine the initial concentration of the two reagents.

$$[H_2]_{initial} = \frac{4.00 \text{ mol}}{2.00 \text{ dm}^3} = 2.00 \text{ mol dm}^{-3}$$

$$[I_2]_{initial} = \frac{6.00 \text{ mol}}{2.00 \text{ dm}^3} = 3.00 \text{ mol dm}^{-3}$$

→ Remember that x represents the amount in mol dm^{-3} of reactant that has reacted at equilibrium.

Initial concentration is the concentration present at the start of the reaction.

Step 3: Set up an equilibrium table.

	H_2	I_2	HI
Initial concentration	2.00 mol dm^{-3}	3.00 mol dm^{-3}	0.00 mol dm^{-3}
Change in concentration	$-x$ mol dm^{-3}	$-x$ mol dm^{-3}	$+2x$ mol dm^{-3}
Equilibrium concentration	$(2.00 - x)$ mol dm^{-3}	$(3.00 - x)$ mol dm^{-3}	$(0.00 + 2x)$ mol dm^{-3}

The change in concentration is determined by the stoichiometry of the reaction. In this case 1 mole of hydrogen reacts with 1 mole of iodine to form 2 moles of hydrogen iodide.

The equilibrium concentrations are given by the sum of the initial concentration term and the change in concentration term for each reactant and product.

Step 4: In order to answer the question we must determine the value of x in order to solve the bottom row of the table. Take the expression for K_c, and replace the terms on the right-hand side with the expressions from the bottom row of the equilibrium table and solve for x:

$$K_c = \frac{[HI]^2}{[H_2][I_2]}$$

$$54 = \frac{(0.00 + 2x)^2}{(2.00 - x)(3.00 - x)}$$

Expand the brackets:

$$54 = \frac{4x^2}{6 - 2x - 3x + x^2}$$

$$54 = \frac{4x^2}{6 - 5x + x^2}$$

$$54(6 - 5x + x^2) = 4x^2$$

$$324 - 270x + 54x^2 = 4x^2$$

$$54x^2 - 4x^2 - 270x + 324 = 0$$

$$50x^2 - 270x + 324 = 0$$

In this case the expression obtained is in the form of a quadratic: $ax^2 + bx + c = 0$ where $a = +50$, $b = -270$, and $c = +324$ and so it is possible to solve for x using the following relationship:

$$x = \frac{-b \pm \sqrt{b^2 - 4ac}}{2a}$$

$$x = \frac{-(-270) \pm \sqrt{(-270)^2 - 4 \times 50 \times 324}}{2 \times 50}$$

$$x = \frac{+270 \pm \sqrt{8100}}{100}$$

$$x = 3.6 \text{ or } 1.8$$

Solving for x in this way will always lead to two solutions for x. However, only one of the possible solutions is sensible. Review the details of the reaction in order to determine the sensible solution. In this case, the reaction started with 2.00 mol dm^{-3} of hydrogen and 3.00 mol dm^{-3} iodine. x represents the concentrations of hydrogen and iodine used up (in mol dm^{-3}) as the reaction achieves equilibrium, hence x must be <2.00. Only one of the two possible solutions is <2.00, so we accept this as the solution and state that $x = 1.8$ mol dm^{-3}.

→ Remember that x represents the number of moles dm^{-3} of reagent that has reacted at equilibrium.

3 CHEMICAL EQUILIBRIUM

→ Check that the answer makes sense by inserting these equilibrium concentrations back into the expression for K_c. Doing so yields a value of 54 for K_c, equal to the value quoted in the question, so the answer is sensible.

Step 5: Determine the composition of the equilibrium mixture:

at equilibrium $[H_2] = (2.00 - x)$ mol dm^{-3} = $(2.00 - 1.8)$ mol dm^{-3} = 0.2 mol dm^{-3};

at equilibrium $[I_2] = (3.00 - x)$ mol dm^{-3} = $(3.00 - 1.8)$ mol dm^{-3} = 1.2 mol dm^{-3};

at equilibrium $[HI] = (0.00 + 2x)$ mol dm^{-3} = $(0.00 + 2(1.8))$ mol dm^{-3} = 3.6 mol dm^{-3}.

Worked example 3.8D

Consider the equilibrium between sulfur dioxide and oxygen to form sulfur trioxide:

$$2\,SO_3\,(g) \rightleftharpoons 2\,SO_2\,(g) + O_2\,(g)$$

→ Note at first glance this question may seem similar to the previous examples but again the mathematical processes used to determine the answer are different.

If sulfur trioxide with a partial pressure of 1.000 atm is allowed to come to equilibrium according to the above reaction and K_p has a value of 4.000×10^{-11} atm, what is the composition of the equilibrium mixture?

Solution

In order to answer the question we have to determine the values of $[SO_3]$, $[SO_2]$, and $[O_2]$ at equilibrium. However, the question only supplies information regarding the starting partial pressure of the reagent and the value of K_p, hence multiple steps are required in order to get to the answer.

Step 1: Begin by writing the expression for K_p, consider the information provided in the question and replace any known terms:

→ Note the only value we know at this stage is the value of K_p.

$$K_p = \frac{(p_{SO_2})^2 \, p_{O_2}}{(p_{SO_3})^2} = 4.000 \times 10^{-11} \text{ atm}$$

→ Remember that x represents the change in partial pressure at equilibrium.

Step 2: Set up an equilibrium table.

	2SO$_3$	2SO$_2$	O$_2$
Initial partial pressure	1.00 atm	0.00 atm	0.00 atm
Change in partial pressure	$-x$ atm	$+x$ atm	$+0.5x$ atm
Equilibrium partial pressure	$(1.00 - x)$ atm	$(0.00 + x)$ atm	$(0.00 + 0.5x)$ atm

Step 3: In order to answer the question we must determine the value of x in order to solve the bottom row of the table. Take the expression for K_p, and replace the terms on the right-hand side with the expressions from the bottom row of the equilibrium table and solve for x:

$$K_p = \frac{(p_{SO_2})^2 \, p_{O_2}}{(p_{SO_3})^2}$$

→ Note that both sides of the expression have the same units, atm. However, the units have been omitted from the calculation to aid clarity.

$$4.000 \times 10^{-11} = \frac{(0.00 + x)^2 (0.00 + 0.5x)}{(1 - x)^2}$$

$$4.000 \times 10^{-11} = \frac{x^2 (0.5x)}{(1 - x)^2}$$

$$4.000 \times 10^{-11} = \frac{0.5x^3}{(1 - x)^2}$$

At this stage it can be seen that a cubic term has been introduced. Consequently, it will not be possible to solve for x using the quadratic approach as in the previous example. However, x can be solved iteratively. Start by assuming that x is very small; the denominator term then becomes equal to 1, and this simplifies the expression allowing x to be determined algebraically.

$$4.000 \times 10^{-11} = 0.5x^3$$

$$\frac{4.000 \times 10^{-11}}{0.5} = x^3$$

$$8.000 \times 10^{-11} = x^3$$

$$\sqrt[3]{8.000 \times 10^{-11}} = x$$

$$4.309 \times 10^{-4} = x$$

In this case x, the change in partial pressure, does appear to be small in comparison to the initial partial pressure of the reagent (1 atm): the change is ~0.04%. So the assumption seems valid. But check by inserting this value of x into the denominator of the expression for K_p and determining a value for numerator value of x. Note this process should be repeated until the denominator and numerator values for x are the same, indicating that the solution is valid.

$$4.000 \times 10^{-11} = \frac{0.5x^3}{(1-x)^2}$$

$$4.000 \times 10^{-11} = \frac{0.5x^3}{\left(1 - 4.309 \times 10^{-4}\right)^2}$$

$$4.000 \times 10^{-11} \times \left(1 - 4.309 \times 10^{-4}\right)^2 = 0.5x^3$$

$$\sqrt[3]{\frac{4.000 \times 10^{-11} \times \left(1 - 4.309 \times 10^{-4}\right)^2}{0.5}} = x$$

$$x = 4.308 \times 10^{-4}$$

→ Remember that x represents the change in partial pressure.

The value for x remains the same so the solution is acceptable: $x = 4.308 \times 10^{-4}$ atm.

Step 4: Determine the composition of the equilibrium mixture:

at equilibrium partial pressure of $SO_3 = (1.00 - x)$ atm $= (1.00 - 4.308 \times 10^{-4})$ atm $= 0.9996$ atm;

at equilibrium partial pressure of $SO_2 = (0.00 + x)$ atm $= (0.00 + 4.308 \times 10^{-4})$ atm $= 4.308 \times 10^{-4}$ atm;

at equilibrium partial pressure of $O_2 = (0.00 + 0.5x)$ atm $= (0.00 + 0.5(4.308 \times 10^{-4}))$ atm $= 2.154 \times 10^{-4}$ atm.

→ Check that the answer makes sense by inserting these equilibrium partial pressures back into the expression for K_p. Doing so yields a value very close to the value quoted in the question, so the answer is sensible.

❓ Question 3.6

Consider the equilibrium between hydrogen and iodine to form hydrogen iodide:

$$H_2(g) + I_2(g) \rightleftharpoons 2HI(g)$$

A mixture of hydrogen, iodine, and hydrogen iodide establishes equilibrium at 700 K. The equilibrium concentrations of hydrogen and iodine are each found to be 3.2 mol dm^{-3}; if K_c has a value of 54, what is the equilibrium concentration of hydrogen iodide?

> **Question 3.7**
>
> Consider the equilibrium between hydrogen and iodine to form hydrogen iodide:
>
> $$H_2(g) + I_2(g) \rightleftharpoons 2HI(g)$$
>
> If 6.00 mol of hydrogen are reacted with 8.00 mol of iodine in a 2.00 dm^3 vessel and K_c has a value of 54 what is the composition of the equilibrium mixture?

> **Question 3.8**
>
> Consider the equilibrium where sulfur trioxide dissociates into sulfur dioxide and oxygen:
>
> $$2SO_3(g) \rightleftharpoons 2SO_2(g) + O_2(g)$$
>
> If sulfur trioxide with a partial pressure of 1.50 atm is allowed to come to equilibrium according to the above reaction and K_p has a value of 9.00×10^{-12} atm, what is the composition of the equilibrium mixture?

3.9 Solubility

The solubility, s, of a compound is the maximum amount in moles per dm^3 that can be dissolved in a given solvent. Compounds may be classified as very soluble, soluble, or sparingly soluble. When the solubility of an ionic compound has been exceeded, equilibrium is established between any solid undissolved material and its ions in solution. The solubility product, K_{sp}, is the equilibrium constant for the process.

➔ When identifying the number and type of ions produced upon dissociation it is very helpful to be familiar with the names and formulae of common ions.

In order to understand the process of dissolution consideration must be given to the number and valency of the ions formed. An ionic compound of general form M_xA_y may dissociate in solution as follows:

$$M_xA_y(s) \rightleftharpoons xM^{a+}(aq) + yA^{b-}(aq)$$

and when this ionic compound is in contact with its saturated solution the thermodynamic equilibrium constant K may be expressed as:

➔ Note that the activity of a pure solid = 1.

$$K = \frac{\left(a_{M^{a+}}\right)^x \left(a_{A^{b-}}\right)^y}{\left(a_{M_xA_y}\right)} = \left(a_{M^{a+}}\right)^x \left(a_{A^{b-}}\right)^y$$

A common approximation to this expression is to consider the ratio of the concentrations of the ions in solution and refer to the constant as the solubility product, K_{sp}:

$$K_{sp} = \left[M^{a+}\right]^x \left[A^{b-}\right]^y$$

As a general rule only the ions produced contribute to K_{sp}. Solids do not form part of the expression because $[M_xA_y]$ is a constant (which is effectively the density of the solid) and therefore is not a variable in the expression.

Worked example 3.9A

Silver iodide, AgI, is a sparingly soluble salt with a K_{sp} value of 8×10^{-17} mol^2 dm^{-6}. What is its solubility?

Solution

1 mole of AgI dissolves in solution to produce 1 mole of univalent silver cations, Ag^+, and 1 mole of iodide anions, I^-:

$$AgI(s) \rightleftharpoons Ag^+(aq) + I^-(aq)$$

When solid AgI is in contact with its saturated solution, the solubility product, K_{sp}, may be expressed in terms of the concentration of each ion; in this case each concentration term is raised to the power 1 as the molar stoichiometry of the ionic compound is 1:1.

$$K_{sp} = [Ag^+][I^-]$$

The square brackets represent the concentration of each ion in solution which in this case is the same as the solubility of each ion, s:

$$K_{sp} = s \times s = s^2$$

➔ Note that for AgI K_{sp} has units of $mol^2\ dm^{-6}$ and s represents solubility, with units of $mol\ dm^{-3}$.

The question provided the value for K_{sp}, so insert this value into the expression and solve for s:

$$8 \times 10^{-17}\ mol^2\ dm^{-6} = s^2$$

$$\sqrt{8 \times 10^{-17}\ mol^2\ dm^{-6}} = 9 \times 10^{-9}\ mol\ dm^{-3} = s$$

➔ Does the answer make sense? Yes: the answer shows that solubility is low—in the nanomolar range—as expected for a sparingly soluble salt.

The solubility, s, of AgI is therefore $9 \times 10^{-9}\ mol\ dm^{-3}$.

Worked example 3.9B

Nickel hydroxide, $Ni(OH)_2$, is a sparingly soluble salt with a K_{sp} value of $6.5 \times 10^{-18}\ mol^3\ dm^{-9}$, what is its solubility?

Solution

First write out an equation for the dissolution process (1 mole of $Ni(OH)_2$ produces 1 mole of divalent Ni^{2+} ions and 2 moles of univalent OH^- ions):

$$Ni(OH)_2(s) \rightleftharpoons Ni^{2+}(aq) + 2\ OH^-(aq)$$

Then write out an expression for the solubility product, K_{sp}, in terms of the concentration of each ion, in this case $[Ni^{2+}]$ is raised to the power 1 and $[OH^-]$ is raised to the power 2 as the stoichiometry is 1:2.

$$K_{sp} = [Ni^{2+}][OH^-]^2$$

➔ K_{sp} will have units of $mol^3\ dm^{-9}$.

The square brackets represent the concentration of each ion in solution which may also be expressed in terms of the solubility of each ion, s. In this case $[Ni^{2+}]$ may be represented as s, whereas $[OH^-]$ being twice the value of $[Ni^{2+}]$ may be represented as $2s$:

$$K_{sp} = s \times (2s)^2 = 4s^3$$

➔ Take care with the algebra here: this expression becomes $s \times 4s^2 = 4s^3$.

The question provided the value for K_{sp}, so insert this value into the expression and solve for s:

$$6.5 \times 10^{-18}\ mol^3\ dm^{-9} = 4s^3$$

$$\sqrt[3]{\frac{6.5 \times 10^{-18}}{4}}\ mol^3\ dm^{-9} = 1.2 \times 10^{-6}\ mol\ dm^{-3} = s$$

➔ Does the answer make sense? Yes: the answer shows that solubility is low—in the micromolar range—as expected for a sparingly soluble salt.

The solubility, s, of $Ni(OH)_2$ is therefore $1.2 \times 10^{-6}\ mol\ dm^{-3}$.

Worked example 3.9C

Calcium hydroxide, Ca(OH)$_2$, is a slightly soluble salt with a solubility of 1.1×10^{-2} mol dm^{-3}. What is its K_{sp} value?

Solution

First write out an equation for the dissolution process assuming complete dissociation of the ions. (1 mole of Ca(OH)$_2$ produces 1 mole of divalent Ca^{2+} ions and 2 moles of univalent OH$^-$ ions.)

$$Ca(OH)_2 \, (s) \rightleftharpoons Ca^{2+} (aq) + 2\, OH^- (aq)$$

Then write out an expression for the solubility product, K_{sp}, in terms of the concentration of each ion. In this case [Ca^{2+}] is raised to the power 1 and [OH$^-$] is raised to the power 2 as the stoichiometry is 1:2.

$$K_{sp} = [Ca^{2+}][OH^-]^2$$

The square brackets represent the concentration of each ion in solution which may also be expressed as the solubility of each ion, s. In this case [Ca^{2+}] may be represented as s, whereas [OH$^-$]—being twice the value of [Ca^{2+}]—may be represented as $2s$

$$K_{sp} = s \times (2s)^2 = 4s^3$$

The question provided the value for s, so insert this value into the expression and solve for K_{sp}:

$$K_{sp} = 4s^3$$

$$K_{sp} = 4(1.1 \times 10^{-2})^3 = 5.3 \times 10^{-6} \text{ mol}^3 \text{ dm}^{-9}$$

> **Note** that here we consider a slightly soluble salt. As the solubility of a compound increases ion–ion interactions in solution can complicate the situation, hence the final answer may be inaccurate. Solubility and solubility product calculations are therefore commonly restricted to sparingly soluble salts only.

> Take care with the algebra here: this expression becomes $s \times 4s^2 = 4s^3$.

> K_{sp} will have units of mol^3 dm^{-9}.

Question 3.9

(a) Silver bromide, AgBr, is a sparingly soluble salt with a K_{sp} value of 7.7×10^{-13} mol^2 dm^{-6}, what is its solubility?

(b) Iron (II) hydroxide, Fe(OH)$_2$, is a sparingly soluble salt with a K_{sp} value of 1.6×10^{-14} mol^3 dm^{-9}, what is its solubility?

(c) Lead sulfide, PbS, is a sparingly soluble salt with a solubility of 1.8×10^{-14} mol dm^{-3}, what is its K_{sp} value?

3.10 Acids, bases, and water

The strength of an acid, HA, is commonly related to the degree of its ionization; the greater the degree of ionization the greater the strength of the acid. Strong acids are completely ionized in solution whilst weak acids are only partially ionized:

$$HA \, (aq) + H_2O \, (l) \rightleftharpoons A^- (aq) + H_3O^+ (aq)$$

and the thermodynamic equilibrium constant K may be expressed as:

$$K = \frac{(a_{A^-})(a_{H_3O^+})}{(a_{HA})(a_{H_2O})} = \frac{(a_{A^-})(a_{H_3O^+})}{(a_{HA})}$$

> Note that the activity of a pure liquid = 1.

A common approximation to this is to consider the ratio of the concentrations instead and to refer to the constant as the acid dissociation constant, K_a:

$$K_a = \frac{[H_3O^+][A^-]}{[HA]}$$

> Remember that pure liquids do not appear in equilibrium constant expressions. Hence K_a will have units of mol dm^{-3}.

Stronger acids have larger K_a values while weaker acids have smaller K_a values. To aid comparison, however, a log scale is often used where:

$$pK_a = -\log_{10} K_a$$

> Note that the p here means $-\log_{10}$. This relationship is analogous with that between pH and $[H_3O^+]$.

Acids donate protons; as such, it is common to express the concentrations of H_3O^+ ions on a log scale referred to as the pH scale:

$$pH = -\log_{10}[H_3O^+]$$

The same approach can be applied to a base. The strength of a base, B, is commonly related to the degree of its ionization:

$$B\,(aq) + H_2O\,(l) \rightleftharpoons BH^+(aq) + OH^-(aq)$$

and the thermodynamic equilibrium constant K may be expressed as:

$$K = \frac{(a_{BH^+})(a_{OH^-})}{(a_B)(a_{H_2O})} = \frac{(a_{BH^+})(a_{OH^-})}{(a_B)}$$

> Note that the activity of a pure liquid = 1.

A common approximation to this is to consider the ratio of the concentrations instead and to refer to the constant as the base ionization constant, K_b:

$$K_b = \frac{[BH^+][OH^-]}{[B]}$$

> Again, remember that pure liquids do not appear in equilibrium constant expressions. Hence K_b will have units of mol dm^{-3}.

Stronger bases have larger K_b values while weaker bases have smaller K_b values. Again a log scale is often used to aid comparison where:

$$pK_b = -\log_{10} K_b$$

Bases accept protons; in an alkaline solution it is convenient to express the concentrations of OH^- ions on a log scale referred to as the pOH scale:

$$pOH = -\log_{10}[OH^-]$$

Remember that strong acids and strong bases are fully ionized and so pK_a and pK_b values are more commonly used to describe the degree of ionization of weak acids and weak bases. In contrast determining pH and pOH for both weak and strong acids is common. Water is amphiprotic which means that it is able to act as both an acid and a base:

$$2\,H_2O(l) \rightleftharpoons H_3O^+(aq) + OH^-(aq)$$

The equilibrium constant for this process is given by:

$$K_w = [H_3O^+][OH^-]$$

> Note that the units of K_w are mol^2 dm^{-6}.

and again the following relationship is often referred to:

$$pK_w = -\log_{10} K_w$$

At 298 K, $K_w = 1.00 \times 10^{-14}$ mol^2 dm^{-6} and $pK_w = 14.00$.

The following relationships are not defined here but they follow from the above and are often used in calculations:

$$K_a \times K_b = K_w$$

$$pK_a + pK_b = pK_w$$

$$pH + pOH = pK_w$$

3 CHEMICAL EQUILIBRIUM

→ Propanoic acid is a weak acid. Note that in this pK_a calculation the number of decimal places in the answer is equal to the number of significant figures in the data.

Worked example 3.10A

If propanoic acid has a K_a value of 1.3×10^{-5} mol dm^{-3} what is its pK_a?

Solution

$$pK_a = -\log_{10} K_a$$

$$pK_a = -\log 1.3 \times 10^{-5}$$

$$pK_a = 4.89$$

Propanoic acid has a pK_a value of 4.89.

Worked example 3.10B

→ Phenol is a very weak acid.

If phenol has a pK_a of 9.99 at 298 K what is its pK_b value?

Solution

$$pK_a + pK_b = pK_w$$

$$pK_b = pK_w - pK_a$$

At 298 K, pK_w = 14.00 so:

$$pK_b = 14.00 - 9.99 = 4.01$$

Phenol has a pK_b value of 4.01.

Worked example 3.10C

→ Nitric acid is a strong acid.

If nitric acid has a pK_a of -1.40 what is its K_a value?

Solution

$$pK_a = -\log_{10} K_a$$

$$-1.40 = -\log_{10} K_a$$

$$10^{1.40} = K_a$$

$$25 = K_a$$

Nitric acid has a K_a value of 25 mol dm^{-3}.

→ Note that the antilog function is simply the inverse of the log function and is often found as a 'second' function on a calculator. This 'second' function can generally be accessed by first pressing the shift button on a calculator followed by the log button.

Worked example 3.10D

→ HCl is a strong acid. Note that in this pH calculation the number of decimal places in the answer is equal to the number of significant figures in the data.

Calculate the pH of a 0.1 mol dm^{-3} HCl aqueous solution.

Solution

$$pH = -\log_{10}[H_3O^+]$$

HCl is a strong acid and completely ionizes in water so $[H_3O^+] = 0.1$ mol dm^{-3} and,

$$pH = -\log_{10}[0.1] = 1.0$$

A 0.1 mol dm^{-3} HCl aqueous solution has a pH of 1.0.

Worked example 3.10E

Calculate the pOH of a 0.25 mol dm^{-3} aqueous NaOH solution.

→ Sodium hydroxide is a strong base. Note that in a pOH calculation the number of decimal places in the answer is equal to the number of significant figures in the data.

Solution

$$pOH = -\log_{10}[OH^-]$$

And $[OH^-] = 0.25$ mol dm^{-3} so:

$$pOH = -\log_{10}[0.25] = 0.60$$

A 0.25 mol dm^{-3} aqueous NaOH solution has a pOH of 0.60.

Worked example 3.10F

Determine the pH of a 0.050 mol dm^{-3} aqueous carboxylic acid (RCOOH) solution with a K_a value of 2.8×10^{-5} mol dm^{-3}.

→ Carboxylic acids are weak acids. Note that here the data has 2 significant figures and so the answer should be quoted to 2 decimal places.

Solution

To determine the pH we must first determine $[H_3O^+]$ so first write an equation representing the ionization of the carboxylic acid:

$$RCOOH(aq) + H_2O(l) \rightleftharpoons H_3O^+(aq) + RCOO^-(aq)$$

Next write an expression representing the acid dissociation constant:

$$K_a = \frac{[H_3O^+][RCOO^-]}{[RCOOH]}$$

→ Note that because water is a pure phase [H$_2$O] is constant and is not part of the expression.

From the dissociation equation $[H_3O^+] = [RCOO^-]$, so:

$$K_a = \frac{[H_3O^+]^2}{[RCOOH]}$$

If we assume that the degree of dissociation is small, [RCOOH] at equilibrium has the same value as the original concentration. The question also provides the value for K_a and so $[H_3O^+]$ can be determined by rearranging the expression to make $[H_3O^+]^2$ the subject and replacing the known terms:

$$[H_3O^+]^2 = K_a[RCOOH]$$

$$[H_3O^+]^2 = (2.8 \times 10^{-5} \text{ mol dm}^{-3}) \times (0.050 \text{ mol dm}^{-3})$$

$$[H_3O^+] = \sqrt{(2.8 \times 10^{-5} \text{ mol dm}^{-3}) \times (0.050 \text{ mol dm}^{-3})}$$

$$[H_3O^+] = 1.2 \times 10^{-3} \text{ mol dm}^{-3}$$

Now we should check the validity of our assumption that the degree of dissociation is small by comparing $[H_3O^+]$ with [RCOOH].

$[H_3O^+] = 1.2 \times 10^{-3}$ mol dm^{-3} and $[RCOOH] = 0.050$ mol dm^{-3} so $[H_3O^+] \ll [RCOOH]$ and the assumption is valid, thus pH can be determined using:

$$pH = -\log_{10}[H_3O^+]$$

$$pH = -\log_{10}[1.2 \times 10^{-3}] = 2.92$$

A 0.050 mol dm^{-3} aqueous carboxylic acid (RCOOH) solution with a K_a value of 2.8×10^{-5} mol dm^{-3} has a pH of 2.92.

> ### Question 3.10
>
> Calculate the pK_a of the following substances:
>
> (a) Hydriodic acid with a K_a value of 1×10^{10} mol dm^{-3};
>
> (b) Carbonic acid with a K_a value of 4.5×10^{-7} mol dm^{-3}.

> ### Question 3.11
>
> Calculate the K_a of the following substances:
>
> (a) Ethanoic acid with a pK_a of 4.76;
>
> (b) Hydrobromic acid has a pK_a of –9.0.

> ### Question 3.12
>
> Calculate the pH and pOH of these solutions at 298 K:
>
> (a) 0.022 mol dm^{-3} HCl (aq);
>
> (b) 0.048 mol dm^{-3} KOH (aq).

> ### Question 3.13
>
> Determine the pH of a 0.17 mol dm^{-3} aqueous carboxylic acid (RCOOH) solution with a K_a value of 4.1×10^{-5} mol dm^{-3}.

3.11 Ligand substitution reactions

→ Note that a Lewis acid is defined as an electron pair acceptor and a Lewis base as an electron pair donor. The ligand is the atom or molecule that coordinates to the metal ion.

d-Block metal ions acting as Lewis acids coordinate with ligands acting as Lewis bases to form coordination complexes. Each complex can subsequently undergo a ligand substitution reaction. Note that several different geometries of complex can be formed. For example, consider hexaaquametal ions in solutions, $[M(H_2O)_6]^{2+}$; the aqua ligands coordinated to the central metal atom can be replaced by ammine ligands:

$$[M(H_2O)_6]^{2+} (aq) + 6\,NH_3 (aq) \rightarrow [M(NH_3)_6]^{2+} (aq) + 6\,H_2O\,(l)$$

3.11 LIGAND SUBSTITUTION REACTIONS

This ligand substitution reaction can be broken down into six successive steps:

$$[M(H_2O)_6]^{2+}(aq) + NH_3(aq) \rightarrow [M(NH_3)(H_2O)_5]^{2+}(aq) + H_2O(l) \quad \text{Step 1}$$

$$[M(NH_3)(H_2O)_5]^{2+}(aq) + NH_3(aq) \rightarrow [M(NH_3)_2(H_2O)_4]^{2+}(aq) + H_2O(l) \quad \text{Step 2}$$

$$[M(NH_3)_2(H_2O)_4]^{2+}(aq) + NH_3(aq) \rightarrow [M(NH_3)_3(H_2O)_3]^{2+}(aq) + H_2O(l) \quad \text{Step 3}$$

$$[M(NH_3)_3(H_2O)_3]^{2+}(aq) + NH_3(aq) \rightarrow [M(NH_3)_4(H_2O)_2]^{2+}(aq) + H_2O(l) \quad \text{Step 4}$$

$$[M(NH_3)_4(H_2O)_2]^{2+}(aq) + NH_3(aq) \rightarrow [M(NH_3)_5(H_2O)]^{2+}(aq) + H_2O(l) \quad \text{Step 5}$$

$$[M(NH_3)_5(H_2O)]^{2+}(aq) + NH_3(aq) \rightarrow [M(NH_3)_6]^{2+}(aq) + H_2O(l) \quad \text{Step 6}$$

Each step has an equilibrium constant known as a stability constant, K, (in this case K_1, K_2, K_3, K_4, K_5, and K_6 respectively). It is commonly found that the stability constant decreases with each successive substitution indicating that at each step the reaction becomes less likely to occur.

The formation constant for the fully substituted compound is referred to as β_n, where n represents the total number of ligands or steps. In this case the formation constant for $[M(NH_3)_6]^{2+}$—an octahedral complex, β_6—is given by the product of all six stability constants:

$$\beta_6 = K_1 \times K_2 \times K_3 \times K_4 \times K_5 \times K_6$$

Ligands can contain varying numbers of donor atoms and are named accordingly—for example, monodentate (contain one donor atom), bidentate (contain two donor atoms) or tridentate (contain three donor atoms)—and the shapes of d-block complexes can be quite varied (some example geometries are tetrahedral, square planar, and octahedral). The units of the formation and stability constants will vary according to the nature of the reaction.

Worked example 3.11A

The addition of ammonia to an aqueous solution of nickel ions may result in substitution of ammine ligands for the aqua ligands. Give an equation that represents the replacement of each aqua ligand in the hexaaquanickel (II) complex ion by an ammine ligand, and write down expressions for the stability constants, K_1–K_6, for the reactions.

Solution

$$[Ni(H_2O)_6]^{2+}(aq) + NH_3(aq) \rightarrow [Ni(NH_3)(H_2O)_5]^{2+}(aq) + H_2O(l) \quad \text{Step 1}$$

$$K_1 = \frac{[[Ni(NH_3)(H_2O)_5]^{2+}]}{[[Ni(H_2O)_6]^{2+}][NH_3]}$$

$$[Ni(NH_3)(H_2O)_5]^{2+}(aq) + NH_3(aq) \rightarrow [Ni(NH_3)_2(H_2O)_4]^{2+}(aq) + H_2O(l) \quad \text{Step 2}$$

$$K_2 = \frac{[[Ni(NH_3)_2(H_2O)_4]^{2+}]}{[[Ni(NH_3)(H_2O)_5]^{2+}][NH_3]}$$

$$[Ni(NH_3)_2(H_2O)_4]^{2+}(aq) + NH_3(aq) \rightarrow [Ni(NH_3)_3(H_2O)_3]^{2+}(aq) + H_2O(l) \quad \text{Step 3}$$

$$K_3 = \frac{[[Ni(NH_3)_3(H_2O)_3]^{2+}]}{[[Ni(NH_3)_2(H_2O)_4]^{2+}][NH_3]}$$

$$[Ni(NH_3)_3(H_2O)_3]^{2+}(aq) + NH_3(aq) \rightarrow [Ni(NH_3)_4(H_2O)_2]^{2+}(aq) + H_2O(l) \quad \text{Step 4}$$

$$K_4 = \frac{[[Ni(NH_3)_4(H_2O)_2]^{2+}]}{[[Ni(NH_3)_3(H_2O)_3]^{2+}][NH_3]}$$

> Check that equations are balanced and that products are placed as the numerator and reactants are placed as the denominator in stability constant expressions. Remember that pure liquids do not appear in equilibrium constant expressions, so H_2O (l) is excluded and the units for each stability constant are $mol^{-1}\ dm^3$.

$[Ni(NH_3)_4(H_2O)_2]^{2+}(aq) + NH_3(aq) \rightarrow [Ni(NH_3)_5(H_2O)]^{2+}(aq) + H_2O(l)$ Step 5

$$K_5 = \frac{[[Ni(NH_3)_5(H_2O)]^{2+}]}{[[Ni(NH_3)_4(H_2O)_2]^{2+}][NH_3]}$$

$[Ni(NH_3)_5(H_2O)]^{2+}(aq) + NH_3(aq) \rightarrow [Ni(NH_3)_6]^{2+}(aq) + H_2O(l)$ Step 6

$$K_6 = \frac{[[Ni(NH_3)_6]^{2+}]}{[[Ni(NH_3)_5(H_2O)]^{2+}][NH_3]}$$

Worked example 3.11B

For the reaction between $[M(H_2O)_6^{2+}]$ and a bidentate ligand, L in which all six water molecules are replaced, three stability constants, $K_1 = 140\ mol^{-1}\ dm^3$, $K_2 = 47\ mol^{-1}\ dm^3$, and $K_3 = 8\ mol^{-1}\ dm^3$ have been determined. Write balanced equations for the three substitution reactions and determine a value for the formation constant β_3.

> Monodentate ligands have 1 donor atom; bidentate ligands have 2 donor atoms.

Solution

$[M(H_2O)_6]^{2+}(aq) + L(aq) \rightarrow [M(L)(H_2O)_4]^{2+}(aq) + 2H_2O(l)$ Step 1

$[M(L)(H_2O)_4]^{2+}(aq) + L(aq) \rightarrow [M(L)_2(H_2O)_2]^{2+}(aq) + 2H_2O(l)$ Step 2

$[M(L)_2(H_2O)_2]^{2+}(aq) + L(aq) \rightarrow [M(L)_3]^{2+}(aq) + 2H_2O(l)$ Step 3

The ligand is bidentate (each ligand contains two donor atoms and replaces two water molecules) and the relationship between the stability constants and the formation constant, is: $\beta_3 = K_1 \times K_2 \times K_3 = 140 \times 47 \times 8 = 52.6 \times 10^3\ mol^{-3}\ dm^9$.

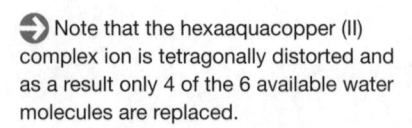

> **? Question 3.14**
>
> The addition of ammonia to an aqueous solution of copper ions may result in substitution of four ammine ligands for the aqua ligands. Give an equation that represents the replacement of each aqua ligand in the hexaaquacopper (II) complex ion by an ammine ligand, and write down expressions for the stability constants, K_1–K_4, for the reactions.

> Note that the hexaaquacopper (II) complex ion is tetragonally distorted and as a result only 4 of the 6 available water molecules are replaced.

Turn to the Synoptic questions section on page 172 to attempt questions that encourage you to draw on concepts and problem-solving strategies from several topics within a given chapter to come to a final answer.

Final answers to numerical questions appear at the end of the book, and full worked solutions appear on the book's website, where you can also find a set of bonus questions for each chapter. Go to www.oxfordtextbooks.co.uk/orc/chemworkbooks/.

4 Phase equilibrium

A phase is a form of matter (gas, liquid, solid, or supercritical fluid) that is uniform in its chemical composition and its physical state. Note that the term 'vapour' is often used to describe the gaseous phase of substances that are liquids or solids at room temperature and the term 'vapour pressure' is the equilibrium partial pressure exerted by a liquid in a sealed container at a fixed temperature. The following section will cover the phase behaviour and phase transitions of one-component systems and also the phase behaviour of two-component mixtures including the phenomena of colligative properties. We recommend you carry out some background reading to review the underlying theory if required.

4.1 Gaseous, liquid, and solid phases

Gaseous systems The atoms and molecules in a gaseous system freely move and expand to fill the available space. The behaviour of a gaseous system can be described using pressure (p), Volume (V), Temperature (T), and the amount of gas (n); some useful relationships are:

At constant temperature and for a fixed amount of gas	$V \propto \dfrac{1}{p}$	$p_1 V_1 = p_2 V_2$	Boyle's law
At constant pressure and for a fixed amount of gas	$V \propto T$	$\dfrac{V_1}{T_1} = \dfrac{V_2}{T_2}$	Charles's law
At constant temperature and pressure	$V \propto n$	$\dfrac{V}{n} = \text{constant}$	Avogadro's law

These relationships combine to give:

$$V \propto \frac{nT}{p}$$

This rearranges to give:

$$pV \propto nT$$

and if we replace the proportionality symbol with a constant we obtain the ideal gas equation:

$$pV = nRT$$

where:

- p is the pressure (measured in Pa).
- V is the volume (measured in m^3).
- n is the amount in moles of gas.
- R is the gas constant (8.314 J K^{-1} mol^{-1}).
- T is the temperature (in K).

➔ When the ideal gas equation is used the quantities must be expressed in SI units.

For a fixed amount of gas then the following relationship is useful in terms of monitoring the effect of any change in conditions:

$$\frac{p_1 V_1}{T_1} = \frac{p_2 V_2}{T_2}$$

where:

- p_1 is the pressure under condition 1.
- V_1 is the volume under condition 1.
- T_1 is the temperature under condition 1.
- p_2 is the pressure under condition 2.
- V_2 is the volume under condition 2.
- T_2 is the temperature under condition 2.

→ When this equation is used temperature must be in kelvin but the units of pressure and volume can be non-SI units as they cancel out. If they are to cancel, however, they must be the same non-SI units.

In an ideal gas mixture, according to **Dalton's law**, the total pressure exerted is the sum of the partial pressures of each individual gas. For example, if a gas mixture comprises three individual gases, A, B, and C, then the total pressure exerted by the mixture is:

$$p_{Total} = p_A + p_B + p_C$$

where:

- p_{Total} is the total pressure exerted.
- p_A is the partial pressure of gas A.
- p_B is the partial pressure of gas B.
- p_C is the partial pressure of gas C.

(All pressure measurements are in pascal.)

The proportion of each gas in a mixture is described in terms of the mole fraction of each individual gas. For example, consider the proportion of gas A present in a mixture; this may be represented as:

$$x_A = \frac{n_A}{n_{Total}}$$

where:

- x_A is the mole fraction of gas A (dimensionless).
- n_A is the amount in moles of gas A (measured in moles).
- n_{Total} is the amount in moles of gas (measured in moles).

The proportion of other gases in the mixture is represented in a similar fashion, so for a mixture that also contains gases B and C, the proportion of these gases present is given by:

$$x_B = \frac{n_B}{n_{Total}}$$

and

$$x_C = \frac{n_C}{n_{Total}}$$

It should be noted that the sum of all mole fractions present must total 1.

The partial pressure of each gas present is related to its mole fraction as follows:

$$p_A = x_A p_{Total}$$

where:

- p_A is the partial pressure of gas A (measured in pascal).
- x_A is the mole fraction of gas A (dimensionless).
- p_{Total} is the total pressure exerted (measured in pascal).

Variations in Gibbs energy for gaseous systems

The term 'ideal gas' refers to a system where the gaseous molecules experience no intermolecular forces; on the whole this situation adequately describes real gases (apart from at high pressures or at temperatures low enough to liquefy the gas). The molar Gibbs energy of an ideal gas depends upon the pressure:

$$G_m = G_m^\ominus + RT \ln \frac{p}{p^\ominus}$$

where:

- G_m is the molar Gibbs energy (measured in J mol^{-1}).
- $G_m^\ominus$ is the standard molar Gibbs energy (measured in J mol^{-1}).
- R is the gas constant (8.314 J K^{-1} mol^{-1}).
- T is temperature (measured in kelvin).
- p is the pressure (measured in bar).
- $p^\ominus$ is the standard pressure (1 bar).

Chemical potential of gaseous systems

For a mixture of gases p can be thought of as the partial pressure of the gas; G_m is then the partial molar Gibbs energy or the chemical potential. The chemical potential of a substance in a mixture is thus the contribution that substance makes to the total Gibbs energy of the mixture:

$$\mu = \mu^\ominus + RT \ln \frac{p}{p^\ominus}$$

> The partial molar Gibbs energy is the contribution that a substance makes to the total molar Gibbs energy of a mixture.

where, for each component:

- μ is the chemical potential of the gaseous component (measured in J mol^{-1}).
- $\mu^\ominus$ is the standard chemical potential of the gaseous component (measured in J mol^{-1}).
- R is the gas constant (8.314 J K^{-1} mol^{-1}).
- T is temperature (measured in kelvin).
- p is the pressure (measured in bar).
- $p^\ominus$ is the standard pressure (1 bar).

If p is defined as the pressure relative to $p^\ominus$, this simply means using a value for p with units of bar for the equation to become:

$$\mu = \mu^\ominus + RT \ln p$$

Real gases

If it is necessary to consider a non-ideal, or real, system then the pressure is replaced with an effective pressure called the fugacity:

$$G_m = G_m^\ominus + RT \ln \frac{f}{p^\ominus}$$

where:

- G_m is the molar Gibbs energy (measured in J mol^{-1}).
- $G_m^\ominus$ is the standard molar Gibbs energy (measured in J mol^{-1}).
- R is the gas constant (8.314 J K^{-1} mol^{-1}).
- T is temperature (measured in kelvin).
- f is the fugacity (measured in bar).
- $p^\ominus$ is the standard pressure (1 bar).

The fugacity is related to pressure as follows:

$$f = \phi p$$

where ϕ is the fugacity coefficient (dimensionless) and p is the pressure (measured in bar).

For an ideal gas the fugacity coefficient has a value of 1. In real systems where the gas molecules attract strongly then the fugacity coefficient has a value <1; f then has a smaller value than p, and the molar Gibbs energy is less than that of an ideal gas under the same conditions. In real systems where the gas molecules repel strongly then the fugacity coefficient has a value >1 and f has a larger value than p.

Liquid phases

Liquids occupy a fixed volume at a given temperature and pressure and take on the shape of the container. Gravitational forces mean the liquid occupies the lower portion of the container and that a well-defined surface exists. The pressure of the vapour in equilibrium with the liquid is called the vapour pressure.

Chemical potential of a liquid phase

An ideal solution is one in which the molecular interactions are all the same. The chemical potential of liquid A is:

$$\mu_A = \mu_A^\ominus + RT \ln x_A$$

where:

- μ_A is the chemical potential of liquid A (measured in J mol^{-1}).
- $\mu_A^\ominus$ is the chemical potential of pure A (measured in J mol^{-1}).
- R is the gas constant (8.314 J K^{-1} mol^{-1}).
- T is temperature (measured in kelvin).
- x_A is the mole fraction of A (dimensionless).

Real solutions

Not all solutions behave ideally. So the equation is modified for each component. For example, the equation is modified for component A as follows:

$$\mu_A = \mu_A^\ominus + RT \ln a_A$$

Where a_A is the activity of A (dimensionless).

Deviations away from ideality occur when the molecules in solutions interact with each other differently; this deviation is often represented in terms of the activity of the species:

> Molality refers to the amount of solute divided by the mass of the solvent; molarity refers to the amount of solute divided by the volume of the solvent.

$$a_A = \gamma_A \frac{b_A}{b^\ominus}$$

where:

- a_A is the activity of A (dimensionless).
- γ_A is the activity coefficient of A (dimensionless).
- b_A is the molality of A (in mol kg^{-1}).
- $b^\ominus$ is 1 mol kg^{-1}.

> A colligative property is one which depends only on the number of solute particles present and not their identity.

In a solution the activity coefficient of the solvent can be determined by measuring a colligative property (boiling point elevation, freezing point depression, or lowering of vapour pressure above a solution). The activity coefficient of the solute is determined from that of the solvent using the Gibbs–Duhem equation:

$$x_A d\mu_A + x_B d\mu_B = 0$$

where:

- x_A is the mole fraction of A (dimensionless).
- $d\mu_A$ is the change of chemical potential of component A (measured in J mol⁻¹).
- x_B is the mole fraction of B (dimensionless).
- $d\mu_B$ is the change of chemical potential of component B (measured in J mol⁻¹).

Thus:

$$d\mu_B = -\frac{x_A d\mu_A}{x_B}$$

Solid phases

Solid phases occupy a fixed volume at a given temperature and pressure and have their own shape and form. Mixtures of solids do exist—for example, different forms of steel are a mixture of different metals.

Most substances form a range of phases and the standard state of a substance is that which exists at exactly 1 bar. (Note that temperature is not part of the definition but by convention data is reported at 298.15 K or 25.00 °C.)

Worked example 4.1A

Calculate the pressure exerted by 8.0 g of oxygen in a 500 cm³ container at 25 °C.

Solution

Assume the gas behaves ideally and use:

$$pV = nRT$$

where:

- p is the pressure (measured in Pa).
- V is the volume (measured in m³).
- n is the amount in moles of gas.
- R is the gas constant (8.314 J K⁻¹ mol⁻¹).
- T is the temperature (in kelvin).

Consider the units detailed above and note that several unit changes are required in order to use the data provided in this equation.

First consider n, which represents the amount of gas in moles. In the question the mass of oxygen is given in g so this must be converted to moles. This is achieved by dividing the mass of oxygen (8.0 g) with the molar mass (M) of oxygen (32 g mol⁻¹):

$$n = \frac{m}{M} = \frac{8.0 \text{ g}}{32 \text{ g mol}^{-1}} = 0.25 \text{ mol}$$

Second consider T, which represents the temperature. In the question the temperature is given in °C and this must be converted to kelvin:

0.00 °C = 273.15 K

So:

25.00 °C = 25.00 + 273.15 = 298.15 K

Third consider V, which represents the volume. In the question the volume is given in cm³ and this must be converted to m³:

1 cm³ is 1 × 10⁻⁶ m³

So:

> There are 100 cm × 100 cm × 100 cm = 1 000 000 cm³ in 1 m³.

500 cm³ is 500×10^{-6} m³

The ideal gas equation can then be rearranged to make pressure the subject:

$$p = \frac{nRT}{V}$$

$$p = \frac{0.25 \text{ mol} \times 8.314 \text{ J K}^{-1}\text{ mol}^{-1} \times 298.15 \text{ K}}{500 \times 10^{-6} \text{ m}^3} = 1.2 \times 10^6 \text{ J m}^{-3}$$

As 1 J m⁻³ is 1 Pa then the pressure exerted by the gas is 1.2×10^6 Pa.

→ Be sure to leave the numerical calculation until the final stage to avoid rounding errors.

> **Question 4.1**
>
> (a) Calculate the pressure exerted by 6.5 g of nitrogen in a 0.05 dm³ container at 37.00 °C.
> (b) Calculate the volume occupied by 2.00 mole of an ideal gas at standard temperature (0.00 °C) and pressure (1.00 atm).

4.2 One-component systems—phase behaviour

For a given system each phase is stable under a specific set of temperature, T, and pressure, p, conditions, as indicated in a phase diagram (a p vs T graphical representation of the thermodynamic stability of each phase with respect to a given set of conditions); Figure 4.1 shows a typical phase diagram. When interpreting a phase diagram note the following:

- A line on a phase diagram indicates the conditions under which the two phases (found either side of the line) are in equilibrium and crossing such a boundary indicates that a phase change is occurring.
- The intersection point, where three lines meet on a phase diagram is known as a triple point and indicates the conditions under which the three phases of matter (gas, liquid, and solid) are all in equilibrium.
- The critical point (where the phase boundary between gas and liquid stops) indicates the point at which a supercritical fluid forms.

Phase transitions are common phenomena for which the central thermodynamic equation is:

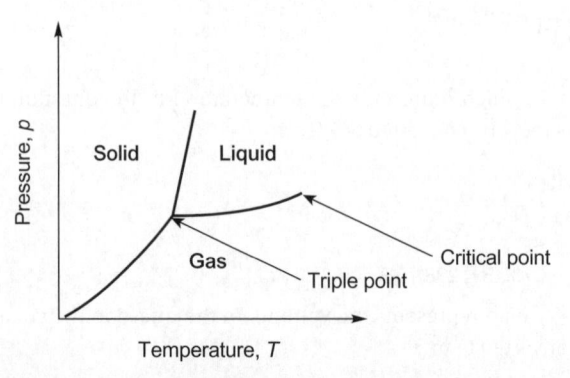

$$G_m = H_m - TS_m$$

Figure 4.1 Typical one-component phase diagram showing the conditions of temperature and pressure at which the solid, liquid, and gas phases are most stable and indicating the position of the triple point, T; the critical point, C; and the equilibria represented by each line.

where:

- G_m is the molar Gibbs energy (J mol^{-1}).
- H_m is the molar enthalpy (J mol^{-1}).
- T is the temperature (K).
- S_m is the molar entropy (J K^{-1} mol^{-1}).

The most stable phase is the one with the lowest Gibbs energy and so understanding how changes in pressure and temperature influence Gibbs energy is important in understanding phase behaviour. Some phase diagrams are complicated by allotropy and polymorphism as each form is considered to be a different phase. Different structural forms of the same element are known as allotropes; some common examples of elements that have allotropes are carbon, phosphorus, sulfur, and tin. Different crystal structures of the same compound are known as polymorphs; some common examples of compounds that have polymorphs are silicon dioxide, calcium carbonate, and ice.

The phase rule can be used to understand the form of a phase diagram:

$$F = C - P + 2$$

F is the number of degrees of freedom (number of intensive variables, such as pressure or temperature, that can be changed without disturbing the number of phases in equilibrium), C is the number of components in the system, and P is the number of phases.

Consider a one-component system:

$$F = 1 - P + 2$$
$$F = 3 - P$$

Table 4.1 summarizes how F varies with P.

Table 4.1 An illustration of how F varies with P for a one-component system.

C	P	F	Implications
1	1	2	A single phase is represented by an 'area' of a phase diagram where p and T can be varied independently.
	2	1	Two phases in equilibrium are represented by a 'line' on a phase diagram where either p or T can be varied but not independently.
	3	0	Three phases in equilibrium are represented by a specific 'point' on a phase diagram where p and T are fixed.

Worked example 4.2A

A pure substance exists as both a solid, a liquid, and a gas phase under different conditions of p and T as indicated in Figure 4.2. Determine how the number of degrees of freedom in the system changes as the system undergoes a phase transition from a liquid to a gas?

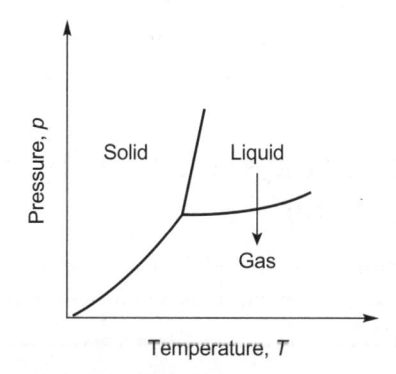

Figure 4.2 Phase diagram for a pure one-component system existing as a solid, a liquid, and a gas phase, indicating a liquid to gas phase change.

Solution

First consider the liquid phase and use the phase rule to determine the number of degrees of freedom, F:

$$F = C - P + 2$$

C is the number of components in the system = 1.
P is the number of phases = 1.
Hence:

$$F = C - P + 2 = 1 - 1 + 2 = 2$$

→ Note in this case C and P both equal 1: there is only one component (the 'pure substance') and one phase (the 'liquid phase').

→ Note that in the liquid part of the phase diagram the number of degrees of freedom, F, = 2. This indicates that there are two intensive variables which are independent of one another; these are 'temperature' and 'pressure'.

Next consider what happens when the pressure is decreased and liquid and gas phases are in equilibrium (indicated by the line). Use the phase rule to determine the number of degrees of freedom, F, when the liquid and gas phases are in equilibrium (that is at the melting point):

$$F = C - P + 2$$

C is the number of components in the system = 1.
P is the number of phases = 2.
Hence:

$$F = C - P + 2 = 1 - 2 + 2 = 1$$

→ Note in this case $C = 1$ and $P = 2$: there is still only one component but it exists as two phases, a 'liquid' and a 'gas'.

→ Note that at the phase transition the number of degrees of freedom, F, = 1. This indicates that at the melting point there is only one independent intensive variable: either temperature or pressure can vary but each is dependent on the other.

Use the phase rule to determine the number of degrees of freedom, F, in the gas phase:

$$F = C - P + 2 .$$

C is the number of components in the system = 1.
P is the number of phases = 1.
Hence:

$$F = C - P + 2 = 1 - 1 + 2 = 2$$

→ Note in this case C and P both equal 1: there is only one component (the 'pure substance') and one phase (the 'single gas phase').

→ Note that in the gas part of the phase diagram the number of degrees of freedom = 2. This indicates that in this part of the phase diagram there are two intensive variables which are independent of one another; these are 'temperature' and 'pressure'.

So in this case the number of degrees of freedom, F, changes from '2' to '1' to '2' as the system changes from a liquid phase to a gas phase.

> **? Question 4.2**
>
> A pure one-component system may exist as a solid, a liquid, or a gas depending on the conditions; determine the number of degrees of freedom, F, when the system comprises:
>
> (a) A pure solid phase.
> (b) A pure liquid phase.
> (c) A pure gas phase.
> (d) An equilibrium mixture of any two phases.
> (e) An equilibrium mixture of the solid, liquid, and gas phases.

4.3 One-component systems—Gibbs energy, enthalpy, and entropy

The most stable phase under a given set of conditions is the one with the lowest molar Gibbs energy, G_m. Molar Gibbs energy, G_m, varies with temperature and pressure. When the temperature is constant the relationship is:

$$dG_m = V_m dp$$

4.3 ONE-COMPONENT SYSTEMS—GIBBS ENERGY, ENTHALPY, AND ENTROPY

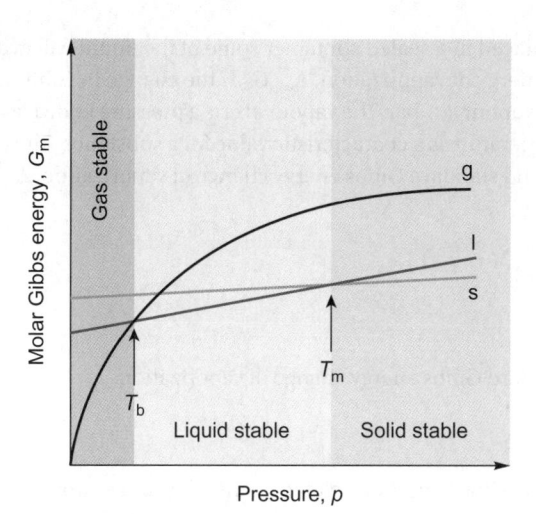

Figure 4.3 Typical variation in molar Gibbs energy, G_m, with pressure, p, for a substance that changes from a gas (g) to a liquid (l) to a solid (s) as the pressure is increased. Boiling point is T_b and melting point is T_m.

where:

- G_m is the molar Gibbs energy.
- V_m is the molar volume.
- p is the pressure.

Figure 4.3 shows the typical variation in molar Gibbs energy, G_m, with pressure, p, for a substance that changes from a gas (g) to a liquid (l) to a solid (s) as the pressure is increased.

When the pressure is constant the relationship is:

$$dG_m = -S_m dT$$

where:

- G_m is the molar Gibbs energy.
- S_m is the molar entropy.
- T is the temperature.

Figure 4.4 shows the typical variation in molar Gibbs energy, G_m, with temperature, T, for a substance that changes from a solid (s) to a liquid (l) to a gas (g) as the temperature is increased.

→ The molar volume is the volume per mole of a substance.

→ Note the slope of the plots in Figure 4.3 equals the molar volume, V_m, of the substance. As expected V_m (s) and V_m (l) are less affected by pressure than V_m (g).

→ Note the slope of the plots in Figure 4.4 equals the molar entropy, S_m, of the substance. As expected S_m (s) < S_m (l) < S_m (g).

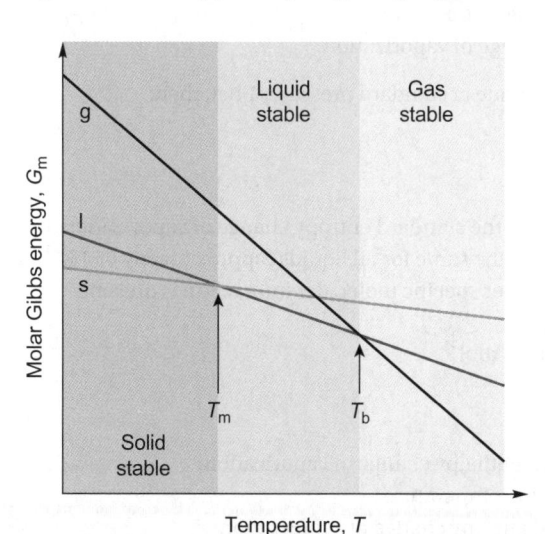

Figure 4.4 Typical variation in molar Gibbs energy, G_m, with temperature, T, for a substance that changes from a solid (s) to a liquid (l) to a gas (g) as the temperature is increased. Melting point is T_m and boiling point is T_b.

When a liquid is placed in a sealed container some of the liquid will evaporate to form a vapour; the standard energy of vaporization, $\Delta_{vap}G^\ominus$ is the change in Gibbs energy when a liquid at 1 bar changes to a vapour at 1 bar. The vapour exerts a pressure known as the vapour pressure which at a given temperature is a characteristic value for a substance. The relationship between vapour pressure, p, and standard Gibbs energy change of vaporization, $\Delta_{vap}G^\ominus$ is:

$$\ln p = -\frac{\Delta_{vap}G^\ominus}{RT}$$

where:

- p is the pressure.
- $\Delta_{vap}G^\ominus$ is the standard Gibbs energy change of vaporization.
- R is the gas constant.
- T is the temperature.

Remember also that Gibbs energy is related to enthalpy and entropy:

$$\Delta_{vap}G^\ominus = \Delta_{vap}H^\ominus - T\Delta_{vap}S^\ominus$$

where:

- $\Delta_{vap}G^\ominus$ is the standard Gibbs energy change of vaporization.
- $\Delta_{vap}H^\ominus$ is the standard enthalpy change of vaporization.
- T is the temperature.
- $\Delta_{vap}S^\ominus$ is the standard entropy change of vaporization.

So another useful equation is:

$$\ln p = -\frac{\Delta_{vap}H^\ominus}{RT} + \frac{\Delta_{vap}S^\ominus}{R}$$

The Gibbs energy change, $\Delta_{vap}G^\ominus$, for the phase transition at constant p and T is zero because the process is at equilibrium. So the enthalpy change, and entropy change of the phase change at a given temperature are related by:

$$\frac{\Delta_{vap}H}{T_{vap}} = \Delta_{vap}S$$

where:

- $\Delta_{vap}H$ is the enthalpy change of vaporization.
- T_{vap} is the boiling temperature.
- $\Delta_{vap}S$ is the entropy change of vaporization.

➔ The normal boiling point of a substance is defined as the temperature at which its vapour pressure is 1 atm (1.013×10^5 Pa). The standard boiling point is defined as the temperature at which its vapour pressure is 1 bar (1.0×10^5 Pa) and strictly speaking this is the number that should be used. However, these numbers are very similar and so this interchange of boiling points has a negligible effect.

If the measurements are made at standard pressure, 1 bar, then:

$$\frac{\Delta_{vap}H^\ominus}{T_{vap}} = \Delta_{vap}S^\ominus$$

Trouton's rule states that the standard entropy change of vaporization at the boiling point of the liquid is approximately the same for all liquids (approximately 85 J K^{-1} mol^{-1}) except when hydrogen bonding or another specific molecular interaction is present:

$$\frac{\Delta_{vap}H^\ominus}{T_{vap}} = \Delta_{vap}S^\ominus \approx 85 \text{ J K}^{-1} \text{ mol}^{-1}$$

where:

- $\Delta_{vap}H^\ominus$ is the standard enthalpy change of vaporization.
- T_b is the normal boiling temperature.
- $\Delta_{vap}S^\ominus$ is the standard entropy change of vaporization.

Worked example 4.3A

Assuming no hydrogen bonding or other specific molecular interactions are present, what is the standard enthalpy of vaporization for a liquid with a normal boiling point of 55.23 °C?

Solution

Take Trouton's rule (which states that the standard entropy change of vaporization, $\Delta_{vap}S^{\ominus}$, at the normal boiling point, T_b, of a liquid (where there is no specific molecular interaction) is approximately 85 J K^{-1} mol^{-1}), insert the boiling point given in the question and rearrange to find the enthalpy of vaporization, $\Delta_{vap}H^{\ominus}$:

$$\frac{\Delta_{vap}H^{\ominus}}{T_b} \approx 85 \text{ J K}^{-1} \text{ mol}^{-1}$$

Remember that the temperature should be in kelvin so the normal boiling point, T_b, should be changed from 55.23 °C to kelvin: (55.23 + 273.15) K = 328.38 K:

→ The units of temperature must be in kelvin in order to get the right answer.

$$\frac{\Delta_{vap}H^{\ominus}}{(328.38 \text{ K})} = 85 \text{ J K}^{-1} \text{ mol}^{-1}$$

$$\Delta_{vap}H^{\ominus} = (85 \text{ J K}^{-1} \text{ mol}^{-1}) \times (328.38 \text{ K}) = 28 \text{ kJ mol}^{-1}$$

The enthalpy of vaporization for a liquid with a boiling point of 55.23 °C is 28 kJ mol^{-1}.

Worked example 4.3B

What is the normal boiling point for a liquid with a standard enthalpy of vaporization of 25 kJ mol^{-1}? Assume there are no hydrogen bonds or other specific molecular interactions in the liquid.

Solution

Take Trouton's rule (which states that the standard entropy change of vaporization, $\Delta_{vap}S^{\ominus}$, at the normal boiling point, T_b, of a liquid (where there is no hydrogen bonding) is approximately 85 J K^{-1} mol^{-1}), insert the standard enthalpy of vaporization, $\Delta_{vap}H^{\ominus}$, given in the question and rearrange to find the normal boiling point, T_b:

→ The enthalpy of vaporization must be in J mol^{-1} in order to get the right answer, note the answer is in kelvin.

$$\frac{\Delta_{vap}H^{\ominus}}{T_b} \approx 85 \text{ J K}^{-1} \text{ mol}^{-1}$$

$$\frac{(25\,000 \text{ J mol}^{-1})}{T_b} = 85 \text{ J K}^{-1} \text{ mol}^{-1}$$

$$T_b = \frac{(25\,000 \text{ J mol}^{-1})}{(85 \text{ J K}^{-1} \text{ mol}^{-1})} = 290 \text{ K}$$

The normal boiling point for a liquid with an enthalpy of vaporization of 25 kJ mol^{-1} is 290 K.

Question 4.3

Determine the standard enthalpy change of vaporization, $\Delta_{vap}H^{\ominus}$, for each of the following liquids assuming there are no specific molecular interactions i.e. no hydrogen bonding.

(a) Liquid A, normal boiling point, $T_b = 310.7$ K.
(b) Liquid B, normal boiling point, $T_b = 47.2$ °C.

4 PHASE EQUILIBRIUM

> **Question 4.4**
>
> Assuming no specific molecular interactions (such as hydrogen bonding) are present, determine the normal boiling point, T_b, for each of the following liquids.
>
> (a) Liquid A, standard enthalpy change of vaporization, $\Delta_{vap}H^\ominus = 26.52$ kJ mol^{-1}.
>
> (b) Liquid B, standard enthalpy change of vaporization, $\Delta_{vap}H^\ominus = 35470$ J mol^{-1}.

4.4 One-component systems—Clapeyron equation

The Clapeyron equation relates the slope of the solid-to-liquid phase boundary on a phase diagram to the change in enthalpy change of fusion, $\Delta_{fus}H$ as follows:

$$\frac{dp}{dT} = \frac{\Delta_{fus}H}{T_m \Delta_{fus}V_m}$$

where:

- p is pressure.
- T is temperature.
- $\Delta_{fus}H$ is the enthalpy change of fusion.
- T_m is the melting point.
- $\Delta_{fus}V_m$ is the change in molar volume.

→ Note the enthalpy of fusion and molar volume do not change very much with temperature hence this is a linear relationship and so a solid-to-liquid phase transition is represented by a straight line in a phase diagram.

→ Note that the slope of the solid-to-liquid phase transition is determined by the change in molar volume. As most compounds expand slightly upon melting (water is an exception) the solid-to-liquid phase transition typically has a positive gradient.

Worked example 4.4A

The normal melting point, T_m, of a substance is 78.02 °C. The respective densities of the solid and liquid phases at this temperature are different giving rise to a molar volume change, ΔV_m of +0.50 cm^3 mol^{-1}. The enthalpy change of fusion, $\Delta_{fus}H$, associated with the transition is +2.5 kJ mol^{-1}. Calculate the melting point of the substance when the pressure is changed to 10 atm.

Solution

First convert some units.

The enthalpy change of fusion, $\Delta_{fus}H$, should be changed from kJ mol^{-1} to J mol^{-1}: 1 kJ mol^{-1} = 1000 J mol^{-1} so +2.5 kJ mol^{-1} = 2.5×10^3 J mol^{-1}.

The normal melting point, T_m, should be changed from 78.02 °C to kelvin: (78.02 + 273.15) K = 351.17 K.

The molar volume change, $\Delta_{fus}V_m$, should be changed from +0.5 cm^3 mol^{-1} to m^3 mol^{-1}: 1 cm^3 = 1×10^{-6} m^3 so +0.5 cm^3 mol^{-1} = 0.5×10^{-6} m^3 mol^{-1}.

Then insert the values into the Clapeyron equation:

$$\frac{dp}{dT} = \frac{\Delta_{fus}H}{T_m \Delta_{fus}V_m}$$

$$\frac{dp}{dT} = \frac{(2.5 \times 10^3 \text{ J mol}^{-1})}{(351.17 \text{ K}) \times (0.50 \times 10^{-6} \text{ m}^3 \text{ mol}^{-1})} = 14.2 \times 10^6 \text{ J m}^{-3} \text{ K}^{-1} = 14.2 \times 10^6 \text{ Pa K}^{-1}$$

→ Note that 1 J m^{-3} = 1 N m m^{-3} = 1 N m^{-2} = 1 Pa and this tells us that a pressure change of 14.2×10^6 Pa will change the melting point by 1 K.

The normal melting point is by definition measured at 1 atm. If the pressure is changed to 10 atm then the change in pressure is (10 atm − 1 atm) = 9 atm and since 1 atm = 1.01×10^5 Pa this corresponds to a pressure change of 909×10^3 Pa.

The melting point temperature change, ΔT, caused by this pressure change, is given by:

$$\Delta T = \frac{(909 \times 10^3 \text{ Pa})}{(14.2 \times 10^6 \text{ Pa K}^{-1})} = 0.064 \text{ K}$$

→ Note the final answer is quoted to 2 significant figures as determined by the data.

So the new melting temperature, T_m, is (351.17 K + 0.064 K) = 351.23 K.

> Note that larger pressure changes will give rise to larger changes in the melting point.

Question 4.5

(a) The normal melting point, T_m, of a substance is 50.02 °C. The respective densities of the solid and liquid phases at this temperature are different giving rise to a molar volume change, ΔV_m of +0.72 cm^3 mol^{-1}. The enthalpy change of vaporization, $\Delta_{fus}H$, associated with the transition is +3.1 kJ mol^{-1}. Calculate the melting point of the substance when the pressure is changed to 5 atm.

(b) A substance is observed to undergo a phase transition (from solid to liquid) at 1 atm pressure and 78.58 °C, with a corresponding molar volume change of +0.35 cm^3 mol^{-1}. The enthalpy change of fusion, $\Delta_{fus}H$, associated with the transition is +4.7 kJ mol^{-1}. Calculate the melting point of the substance when the pressure is changed to 20 atm.

4.5 One-component systems—Clausius–Clapeyron equation

If a transition involves a gas or vapour phase then the Clapeyron equation is modified and becomes the Clausius–Clapeyron equation:

$$\frac{dp}{dT} = p\frac{\Delta_{vap}H}{RT^2}$$

which (away from the critical point) commonly takes the form:

$$\ln\frac{p_1}{p_2} = \frac{\Delta_{vap}H}{R}\left(\frac{1}{T_2} - \frac{1}{T_1}\right)$$

It is interesting to note that the Clausius–Clapeyron equation can take other forms:

$$\ln\frac{p_2}{p_1} = -\frac{\Delta_{vap}H}{R}\left(\frac{1}{T_2} - \frac{1}{T_1}\right)$$

or:

$$\ln\frac{p_2}{p_1} = \frac{\Delta_{vap}H}{R}\left(\frac{1}{T_1} - \frac{1}{T_2}\right)$$

However, all forms yield the same answer.

Depending upon what parameters are known, this equation may be used to determine how phase transition temperatures vary with pressure, and how vapour pressure varies with temperature and either enthalpy of vaporization (as shown here) or, of course, enthalpy of sublimation.

> Note this is not a linear relationship and so a solid-to-gas and a liquid-to-gas transition is represented by a curved line in a phase diagram.

> Any units for pressure can be used as long as they are the same for p_1 and p_2. The units used for temperature must be kelvin.

Worked example 4.5A

The normal boiling point, T_b, of a substance is 88.02 °C. The enthalpy change of vaporization, $\Delta_{vap}H$, associated with the transition is +25.0 kJ mol^{-1}. Calculate the boiling point of the substance when the pressure is changed to 101 Pa.

Solution

In this case the enthalpy change of vaporization and the boiling point at a given pressure are provided so it is possible to use the Clausius–Clapeyron equation to determine the boiling point at a second pressure:

$$\ln\frac{p_1}{p_2} = \frac{\Delta_{vap}H}{R}\left(\frac{1}{T_2} - \frac{1}{T_1}\right)$$

First convert some units.

The enthalpy of vaporization, $\Delta_{vap}H$, should be changed from kJ mol^{-1} to J mol^{-1}: 1 kJ mol^{-1} = 1000 J mol^{-1} so +25.0 kJ mol^{-1} = 25.0 × 10^3 J mol^{-1}.

The normal boiling point, T_b, should be changed from 88.02 °C to kelvin: (88.02 + 273.15) K = 361.17 K.

The initial pressure, p_1, should be changed from atm to Pa: by definition the normal melting point is measured at 1 atm and as 1 atm = 1.01 × 10^5 Pa then p_1 = 1.01 × 10^5 Pa.

Then insert the values into the equation and rearrange to make T_2 the subject:

→ Note that on the left-hand side there is a natural logarithm term. The natural logarithm of a number is unitless so the units of pressure subsequently disappear from the expression.

$$\ln\frac{(1.01\times10^5\,\text{Pa})}{(101\,\text{Pa})} = \frac{(25.0\times10^3\,\text{J mol}^{-1})}{(8.314\,\text{J K}^{-1}\,\text{mol}^{-1})} \times \left(\frac{1}{T_2} - \frac{1}{361.17\,\text{K}}\right)$$

$$6.908 = 3007\,\text{K} \times \left(\frac{1}{T_2} - 2.769\times10^{-3}\,\text{K}^{-1}\right)$$

$$6.908 = \left(\frac{3007\,\text{K}}{T_2}\right) - 8.326$$

$$15.23 = \frac{(3007\,\text{K})}{T_2}$$

$$T_2 = \frac{(3007\,\text{K})}{15.23} = 197.4\,\text{K}$$

→ Does the answer make sense? Yes: when the pressure is reduced the boiling point lowers as expected.

The new boiling point of the substance is 197.4 K or −75.7 °C.

❓ Question 4.6

(a) The normal boiling point, T_b, of a substance is 101.33 °C. The enthalpy change, $\Delta_{vap}H$, associated with the transition is +29.0 kJ mol^{-1}. Calculate the boiling point of the substance when the pressure is changed to 0.500 atm.

(b) A substance is observed to undergo a phase transition (from liquid to gas) at 120.45 °C with a corresponding enthalpy change, $\Delta_{fus}H$, of +34.0 kJ mol^{-1}. Calculate the boiling point of the substance when the pressure is changed to 2.00 atm.

Worked example 4.5B

Water, with an enthalpy change of vaporization, $\Delta_{vap}H$, of 40.65 kJ mol^{-1} and a vapour pressure of 1.000 atm, is held in a sealed container at 25.00 °C. Determine the vapour pressure if the temperature is increased to 99.00 °C.

4.5 ONE-COMPONENT SYSTEMS – CLAUSIUS–CLAPEYRON EQUATION

Solution

Given the vapour pressure, p_1, at the initial temperature, T_1, it is possible to use the Clausius–Clapeyron equation to determine the vapour pressure, p_2, at the second temperature, T_2:

$$\ln\frac{p_1}{p_2} = \frac{\Delta_{vap}H}{R}\left(\frac{1}{T_2} - \frac{1}{T_1}\right)$$

First convert some units.

The enthalpy change of vaporization, $\Delta_{vap}H$, should be changed from kJ mol^{-1} to J mol^{-1}: 1 kJ mol^{-1} = 1000 J mol^{-1} so +40.65 kJ mol^{-1} = 40.65 × 10^3 J mol^{-1}.

T_1, should be changed from 25.00 °C to kelvin: (25.00 + 273.15) K = 298.15 K.
T_2, should be changed from 99.00 °C to kelvin: (99.00 + 273.15) K = 372.15 K.
Then insert the values into the equation:

$$\ln\frac{(1.000\text{ atm})}{p_2} = \left(\frac{(40.65\times 10^3\text{ J mol}^{-1})}{(8.314\text{ J K}^{-1}\text{ mol}^{-1})}\times\left(\frac{1}{372.15\text{ K}} - \frac{1}{298.15\text{ K}}\right)\right)$$

→ Do not solve the right-hand side of this expression just yet so as to avoid introducing a rounding error.

In order to determine p_2 we need to remove the natural logarithm term from the left-hand side of the expression; the rule used is that if $\ln x = y$ then $x = e^y$.

$$\frac{(1.000\text{ atm})}{p_2} = e^{\left(\frac{(40.65\times 10^3\text{ J mol}^{-1})}{(8.314\text{ J K}^{-1}\text{ mol}^{-1})}\times\left(\frac{1}{372.15\text{ K}} - \frac{1}{298.15\text{ K}}\right)\right)}$$

Note that e is a mathematical constant: the base of the natural logarithm with a value of 2.718. On a scientific calculator you will find an antilog, e^x, function which allows the value of the exponential function to be determined:

→ Does the answer make sense? Yes: when the temperature is raised then the vapour pressure inside the sealed container increases as expected.

$p_2 = 26.07$ atm

The new vapour pressure is 26.07 atm.

> **? Question 4.7**
>
> (a) Benzene, with an enthalpy change of vaporization, $\Delta_{vap}H$, of 30.80 kJ mol^{-1} and a vapour pressure of 94.60 torr, is held in a sealed container at 25.00 °C. Determine the vapour pressure if the temperature is increased to 45.00 °C.
>
> (b) Estimate the vapour pressure of water at 50.00 °C, given that the normal boiling point of water is 100.00 °C and that it has an enthalpy change of vaporization, $\Delta_{vap}H$, of 40.65 kJ mol^{-1}.

Worked example 4.5C

The vapour pressure of a substance is 54 torr at 0.00 °C and 345 torr at 25.00 °C. Determine the enthalpy change of vaporization, $\Delta_{vap}H$.

Solution

Given the vapour pressure, p_1, at one temperature, T_1, and the vapour pressure, p_2, at a second temperature, T_2, it is possible to use the Clausius–Clapeyron equation to determine the enthalpy change of vaporization, $\Delta_{vap}H$:

$$\ln\frac{p_1}{p_2} = \frac{\Delta_{vap}H}{R}\left(\frac{1}{T_2} - \frac{1}{T_1}\right)$$

First convert some units.

T_1, should be changed from 0.00 °C to kelvin: (0.00 + 273.15) K = 273.15 K.
T_2, should be changed from 25.00 °C to kelvin: (25.00 + 273.15) K = 298.15 K.

Then insert the values into the equation and rearrange to make the enthalpy of vaporization the subject:

$$\ln\frac{54}{345} = \frac{\Delta_{vap}H}{(8.314\text{ J K}^{-1}\text{mol}^{-1})}\left(\frac{1}{298.15\text{ K}} - \frac{1}{273.15\text{ K}}\right)$$

$$-1.855 = \frac{\Delta_{vap}H}{(8.314\text{ J K}^{-1}\text{mol}^{-1})}\left(-3.07\times10^{-4}\text{ K}^{-1}\right)$$

$$\Delta_{vap}H = 50\text{ kJ mol}^{-1}$$

→ To avoid rounding errors it is best to leave the numerical calculation to the final stage.

The standard enthalpy of vaporization of the substance is 50 kJ mol^{-1}.

> **? Question 4.8**
>
> (a) Calculate the enthalpy change of vaporization for a substance with the following vapour pressures: 760 mm Hg at 85.00 °C and 450 mm Hg at 55.00 °C.
>
> (b) The vapour pressure of a substance is 34.0 torr at 15.00 °C and 410 torr at 65.00 °C. Determine the enthalpy change of vaporization, $\Delta_{vap}H$.

4.6 Two-component mixtures—non-volatile solute plus volatile solvent

The vapour pressure of a solvent is lower in the presence of a non-volatile solute. In fact, according to Raoult's law, the vapour pressure of the solvent is proportional to its mole fraction in solution:

$$p = x_{solvent} \times p_{pure}$$

where:

- p is the vapour pressure of the solvent in solution.
- p_{pure} is the vapour pressure of the pure solvent.
- $x_{solvent}$ is the mole fraction of the solvent.

If the mole fraction of the non-volatile component is known (for example, assume it to be 0.010) then the mole fraction of the solvent is 0.990. If the vapour pressure of the pure solvent is 20.0 torr then the vapour pressure of the solution is:

$$p_{solvent} = 0.990 \times (20.0\text{ torr}) = 19.8\text{ torr}$$

When a non-volatile solute is present in solution the vapour pressure of the solvent is lowered and there is a corresponding lowering of the solvent freezing point, ΔT_f, elevation of the boiling point, ΔT_b, and increase in osmotic pressure, Π, according to the following three equations:

$$\Delta T_f = ik_f m$$
$$\Delta T_b = ik_b m$$
$$\Pi = iRTc$$

where:

- ΔT_f = depression in freezing point (measured in kelvin).
- ΔT_b = elevation of boiling point (measured in kelvin).

4.6 TWO-COMPONENT MIXTURES—NON-VOLATILE SOLUTE PLUS VOLATILE SOLVENT

- Π = osmotic pressure of the solution (measured in units of pressure).
- i = van't Hoff factor (approximately the number of species produced by the solute when in solution; for non-electrolytes this is 1, for MX salts this is 2, for MX_2 salts this is 3).
- k_f = freezing point constant of the solvent (measured in K kg mol^{-1}).
- k_b = boiling point constant of the solvent (measured in K kg mol^{-1}).
- m = molality (measured in mol kg^{-1}).
- c = molarity (measured in mol dm^{-3}).
- R = gas constant (0.0821 dm^3 atm K^{-1} mol^{-1}).
- T = temperature (measured in kelvin).

These properties (depression of freezing point, elevation of boiling point, and osmotic pressure) are called colligative properties because they depend on the amount of material added and not the nature of the material.

> Colligative properties are those that depend only on the number of dissolved particles in solution and not on their nature.

Worked example 4.6A

Determine at what temperature a 0.200 m aqueous solution of sucrose will freeze given that the normal freezing point of water is 0 °C and the freezing point constant, k_f, for water is 1.86 K kg mol^{-1}.

Solution

Take the expression that relates freezing point depression to molality. Sucrose is a non-electrolyte so it has an i value of 1. Insert the other values given.

$$\Delta T_f = i k_f m$$

$$\Delta T_f = 1 \times (1.86 \text{ K kg mol}^{-1}) \times (0.200 \text{ mol kg}^{-1})$$

$$\Delta T_f = 0.372 \text{ K} \approx 0.4 \text{ K}$$

Hence the temperature at which the water freezes is 0 °C − 0.4 °C = −0.4 °C.

> This represents the observed decrease in freezing point; note that a decrease of 0.4 K is the same as a decrease of 0.4 °C.

Question 4.9

(a) What is the freezing point of a 0.070 m aqueous solution of sucrose given that the normal freezing point of water is 0.00 °C and the freezing point constant, k_f, for water is 1.86 K kg mol^{-1}.

(b) Determine at what temperature a 0.10 m solution of aspirin in cyclohexane will freeze given that the normal freezing point of cyclohexane is 6.5 °C and the freezing point constant, k_f, for cyclohexane is 20.1 K kg mol^{-1}.

Worked example 4.6B

Determine at what temperature a 0.10 m aqueous solution of NaCl will boil given that the normal boiling point of water is 100.00 °C and the boiling point constant, k_b, for water is 0.51 K kg mol^{-1}.

Solution

Take the expression that relates boiling point elevation to molality. NaCl is an electrolyte that completely dissociates upon dissolution to produce Na⁺ and Cl⁻ ions so it has an *i* value of 2. Insert the other values given.

$$\Delta T_b = i k_b m$$

$$\Delta T_b = 2 \times (0.51 \text{ K kg mol}^{-1}) \times (0.10 \text{ mol kg}^{-1})$$

$$\Delta T_b = 0.10 \text{ K}$$

> This represents the observed change in boiling point. In this case the temperature increases.

> Note that an increase of 0.1 K is the same as an increase of 0.1 °C.

Hence the temperature at which the solution boils is 100.00 °C + 0.10 °C = 100.10 °C.

Question 4.10

(a) Determine at what temperature a 0.10 m solution of glucose in cyclohexane will boil given that the normal boiling point of cyclohexane is 80.7 °C and the boiling point constant, k_b, for cyclohexane is 2.79 K kg mol⁻¹.

(b) What is the boiling point of a 0.50 m aqueous solution of MgCl₂ given that the normal boiling point of water is 100.0 °C and the boiling point constant, k_b, for water is 0.51 K kg mol⁻¹?

Worked example 4.6C

Determine the osmotic pressure of a 0.20 M aqueous solution of sucrose at 25 °C.

Solution

Take the expression that relates osmotic pressure to molarity. Sucrose is a non-electrolyte so assume that *i* has a value of 1. Insert the other values given remembering to convert temperature to kelvin.

$$\Pi = iRTc$$

$$\Pi = 1 \times (0.0821 \text{ dm}^3 \text{ atm K}^{-1} \text{ mol}^{-1}) \times (298 \text{ K}) \times (0.20 \text{ mol dm}^{-3})$$

$$\Pi = 4.9 \text{ atm}$$

> The units of osmotic pressure will depend upon the units of *R* selected. Here the chosen value of *R* will produce an osmotic pressure with units of atm.

The osmotic pressure is 4.9 atm.

Question 4.11

(a) Determine the osmotic pressure of a 0.10 M aqueous solution of NaCl at 50 °C.

(b) Determine the osmotic pressure of a 1×10^{-4} M aqueous solution of glucose at 37 °C.

Worked example 4.6D

An isolated oil extract (a non-electrolyte) was found to lower the freezing point of camphor: when 0.1 g of the extract was added to 100 g of camphor the freezing point was lowered by

0.500 °C. Calculate the molar mass of the extract given that the freezing point constant for camphor is 39.7 K kg mol^{-1}.

Solution

Take the expression that relates freezing point depression with molality and rearrange to make molality of the solution the subject and insert the values given:

$$\Delta T_f = ik_f m$$

$$m = \frac{\Delta T_f}{ik_f}$$

➔ Assume that $i = 1$ as the extract is a non-electrolyte.

$$m = \frac{(0.500 \text{ K})}{1 \times (39.7 \text{ K kg mol}^{-1})} = 0.0126 \text{ mol kg}^{-1}$$

➔ This is the molality of the solution (that is the amount in moles of extract per kilogram of camphor).

The mass of camphor is 100 g or 0.100 kg so the amount in moles of extract is:

Moles of extract $= (0.100 \text{ kg}) \times (0.0126 \text{ mol kg}^{-1}) = 1.26 \times 10^{-3}$ mol

And the molar mass of the extract is:

Molar mass of extract $= \dfrac{\text{mass of extract}}{\text{moles of extract}} = \dfrac{0.1 \text{ g}}{1.26 \times 10^{-3} \text{ mol}} = 79.4$ g mol^{-1}

➔ To avoid rounding errors it is best to leave the numerical calculation to the final stage.

The molar mass of the extract is 79.4 g mol^{-1}.

❓ Question 4.12

(a) When 0.15 mg of a substance (a non-electrolyte) was added to 100 mg of camphor, the freezing point was lowered by 2.8 °C. Calculate the molar mass of the substance given that the freezing point constant for camphor is 39.7 K kg mol^{-1}.

(b) An oil extract (a non-electrolyte) was found to lower the freezing point of cyclohexane; when 0.20 g of the extract was added to 100 g of cyclohexane the freezing point was lowered by 1.2 °C. Calculate the molar mass of the extract given that the freezing point constant for cyclohexane is 20.0 K kg mol^{-1}.

Worked example 4.6E

2 g of a polymer was dissolved in methylbenzene to produce a 100 ml sample with an osmotic pressure of 0.1 atm at 25 °C. Determine the molar mass of the polymer assuming it is a non-electrolyte.

Solution

Take the expression that relates osmotic pressure to molarity, rearrange to make the molarity of the solution the subject, and insert the values given remembering to convert temperature to kelvin:

$$\Pi = iRTc$$

$$c = \frac{\Pi}{iRT}$$

➔ Assume that $i = 1$.

$$c = \frac{(0.1 \text{ atm})}{1 \times (0.0821 \text{ dm}^3 \text{ atm K}^{-1} \text{ mol}^{-1}) \times (298 \text{ K})} = 4 \times 10^{-3} \text{ mol dm}^{-3}$$

As the sample volume was 100 ml (or 0.1 dm³) it follows that the amount in moles, n, in the sample is:

$$n = (0.1\,\text{dm}^3) \times (4 \times 10^{-3}\,\text{mol}\,\text{dm}^{-3}) = 4 \times 10^{-4}\,\text{mol}$$

The original mass of polymer was 2 g so the molar mass is:

$$\text{Molar mass} = \frac{(2\,\text{g})}{(4 \times 10^{-4}\,\text{mol})} = 5000\,\text{g}\,\text{mol}^{-1}$$

The molar mass of the polymer is 5000 g mol⁻¹.

→ The molar mass of the polymer is determined to be 5000 g mol⁻¹. It is best to leave the numerical calculation to the final stage to avoid rounding errors.

> **Question 4.13**
>
> (a) 5 g of a polymer was dissolved in methylbenzene to produce a 100 ml sample with an osmotic pressure of 0.2 atm at 30 °C. Determine the molar mass of the polymer assuming it is a non-electrolyte.
>
> (b) 3.50 g of a polymer was dissolved in cyclohexane to produce a 200 ml sample with an osmotic pressure of 0.500 atm at 25 °C. Determine the molar mass of the polymer assuming it is a non-electrolyte.

4.7 Two-component mixtures—ideal binary liquid mixtures

An ideal solution is one in which the molecules experience the same intermolecular interactions as in the pure liquids. Consider an ideal mixture of two components A and B in which each component obeys Raoult's law.

The vapour pressure of A, p_A, is given by:

$$p_A = x_{A,\,\text{liquid}} \times p_{A,\,\text{pure}}$$

Where $x_{A,\,\text{liquid}}$ is the mole fraction of A in the liquid mixture and $p_{A,\,\text{pure}}$ is the vapour pressure of pure A.

The vapour pressure of B, p_B, is given by:

$$p_B = x_{B,\,\text{liquid}} \times p_{B,\,\text{pure}}$$

Where $x_{B,\,\text{liquid}}$ is the mole fraction of B in the liquid mixture and $p_{B,\,\text{pure}}$ is the vapour pressure of pure B.

According to Dalton's law the total vapour pressure, p_{Total}, is the sum of the two partial vapour pressures:

$$p_{\text{Total}} = p_A + p_B = (x_{A,\,\text{liquid}} \times p_{A,\,\text{pure}}) + (x_{B,\,\text{liquid}} \times p_{B,\,\text{pure}})$$

Figure 4.5 shows that the total vapour pressure of an ideal binary liquid mixture depends upon its composition and is the sum of the partial pressures of the two components.

The composition of the vapour in equilibrium with the liquid tends to be richer in the more volatile component and is given by:

$$x_{A,\,\text{vapour}} = \frac{p_A}{p_{\text{Total}}} = \frac{p_A}{p_A + p_B} = \frac{x_{A,\,\text{liquid}} \times p_{A,\,\text{pure}}}{(x_{A,\,\text{liquid}} \times p_{A,\,\text{pure}}) + (x_{B,\,\text{liquid}} \times p_{B,\,\text{pure}})}$$

The relationship between the composition of the liquid and the composition of the vapour is illustrated in Figure 4.6.

4.7 TWO-COMPONENT MIXTURES—IDEAL BINARY LIQUID MIXTURES

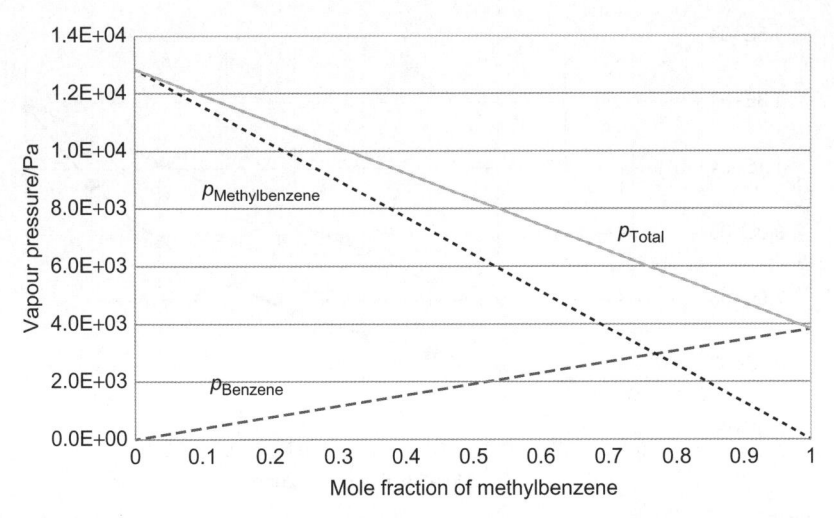

Figure 4.5 The total vapour pressure of an ideal binary liquid mixture depends upon the composition and is the sum of the vapour pressures of the two components. The example shown here is for methylbenzene and benzene (two similar liquids).

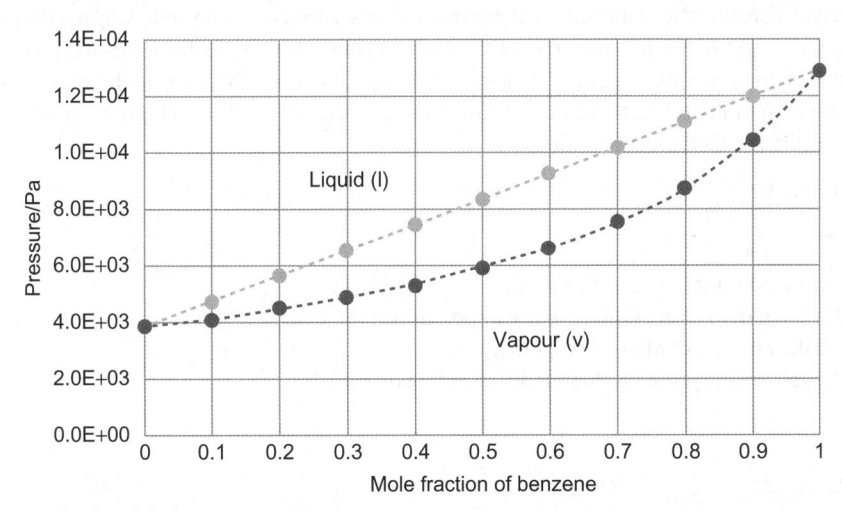

Figure 4.6 Pressure–composition liquid–vapour phase diagram for an ideal solution at fixed temperature, showing an example of methylbenzene and benzene (two similar liquids).

At high pressure (above the upper line) only a liquid phase exists; at low pressure (beneath the lower line) only a vapour phase exists; in the middle region (between the upper and lower lines) both liquid and vapour phases co-exist. Consider Figure 4.6 in more detail: in this middle region the relative amounts of liquid and vapour phases at different pressures may be determined as illustrated in the worked example that follows.

Worked example 4.7A

Look at Figure 4.7 and consider a mixture of overall composition 0.5 mole fraction at a pressure of 7 kPa. At this pressure both liquid and vapour phases co-exist with different compositions.

First, consider the composition of the liquid phase; this is given by following the tie-line L1 towards the left-hand side. Note the mole fraction on the x-axis where the tie-line intersects with the upper line on the plot (approx. 0.36): this is the mole fraction of benzene in the liquid phase.

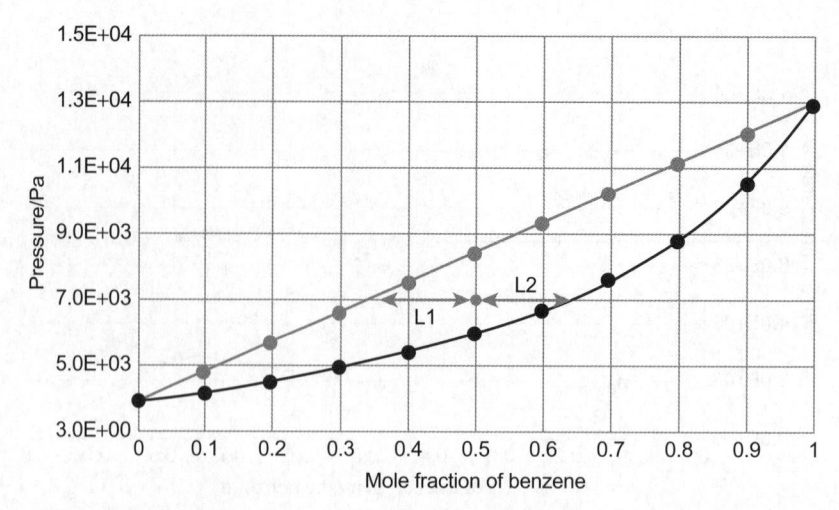

Figure 4.7 Pressure–composition liquid–vapour phase diagram for an ideal solution at fixed temperature. Example plot for methylbenzene and benzene (two similar liquids). L1 and L2 represent tie-lines in the two-phase region.

Second, consider the composition of the vapour phase; this is given by following the tie-line L2 towards the right-hand side. Note the mole fraction on the x-axis where the tie line intersects with the lower line on the plot (approx. 0.64): this is the mole fraction of benzene in the vapour phase.

The length of the tie-lines, L1 and L2, are related to the amount of each phase present. This relationship is called the lever rule:

$$n_1 L1 = n_2 L2$$

where:

- n_1 is the amount of phase 1 (liquid).
- L1 is the length of the tie-line to the upper line on the plot.
- n_2 is the amount of phase 2 (vapour).
- L2 is the length of the tie-line to the lower line on the plot.

or:

$$\frac{n_1}{n_{\text{Total}}} = \frac{L2}{L_{\text{Total}}} \quad \text{and} \quad \frac{n_2}{n_{\text{Total}}} = \frac{L1}{L_{\text{Total}}}$$

where:

- n_1 is the amount of phase 1 (liquid).
- n_{Total} is the total amount of both phases.
- L2 is the length of the tie-line to the lower line on the plot.
- L_{Total} is the total length of both tie-lines.
- n_2 is the amount of phase 2 (vapour).
- L1 is the length of the tie-line to the upper line on the plot.

Note the lever rule can be applied to any two-phase, two-component system.

Worked example 4.7B

Consider Figure 4.7. A closed system composed of 12 moles of benzene and 12 moles of methylbenzene (such that the mole fraction of benzene is therefore 0.5) is initially held at a sufficiently high pressure so that only a liquid phase is present; the pressure is then lowered isothermally to 7 kPa where a liquid and vapour phase are in equilibrium. What are the compositions of these two phases?

Solution

The composition of the liquid phase is read from the diagram: benzene has a mole fraction of 0.36 and so methylbenzene has a mole fraction of 0.64.

Similarly the composition of the vapour phase is read from the diagram: benzene has a mole fraction of 0.64 and so methylbenzene has a mole fraction of 0.36.

The length of the two lines L1 and L2 should be measured and the total amount in moles of benzene and methylbenzene in the liquid phase (n_1) determined using:

$$n_1 = \frac{L2}{L_{Total}} \times n_{Total}$$

$$n_1 = \frac{0.14}{0.28} \times 24$$

$$n_1 = 12$$

How many of these moles are benzene and how many methylbenzene? This is determined as follows:

Moles of benzene in liquid phase = mole fraction of benzene in liquid phase $\times n_1$

Moles of benzene in liquid phase = 0.36×12

Moles of benzene in liquid phase = 4.32

and:

Moles of methylbenzene in liquid phase = mole fraction of methylbenzene in liquid phase $\times n_1$

Moles of methylbenzene in liquid phase = 0.64×12

Moles of methylbenzene in liquid phase = 7.68

Similarly the total amount in moles of benzene and methylbenzene in the vapour phase (n_2) is then determined using:

$$n_2 = \frac{L1}{L_{Total}} \times n_{Total}$$

$$n_2 = \frac{0.14}{0.28} \times 24$$

$$n_2 = 12$$

How many of these moles are benzene and how many methylbenzene? This is determined as follows:

Moles of benzene in vapour phase = mole fraction of benzene in vapour phase $\times n_1$

Moles of benzene in vapour phase = 0.64×12

Moles of benzene in vapour phase = 7.68

Finally check that the total number of moles is correct and as expected: $(4.32 + 7.68 + 7.68 + 4.32) = 24$.

Worked example 4.7C

The vapour pressures of benzene and methylbenzene are 94.6 and 29.1 torr at 25 °C. What is the total vapour pressure of a mixture that contains two moles of benzene and eight moles of methylbenzene?

Solution

Benzene and methylbenzene are similar in molecular structure and have similar chemical interactions. The mixture will therefore behave ideally. The mole fraction of benzene, x_{Benzene}, is:

$$x_{\text{Benzene}} = \frac{\text{Amount in moles of benzene}}{\text{Total amount in moles}} = \frac{2}{10} = 0.2$$

So the partial pressure of benzene is:

$$p_{\text{Benzene}} = x_{\text{Benzene, liquid}} \times p_{\text{Benzene, pure}} = 0.2 \times (94.6 \text{ torr}) = 18.9 \text{ torr}$$

The mole fraction of methylbenzene, $x_{\text{Methylbenzene}}$, is:

$$x_{\text{Methylbenzene}} = \frac{\text{Amount in moles of methylbenzene}}{\text{Total amount in moles}} = \frac{8}{10} = 0.8$$

So the partial pressure of methylbenzene is:

$$p_{\text{Methylbenzene}} = x_{\text{Methylbenzene, liquid}} \times p_{\text{Methylbenzene, pure}} = 0.8 \times (29.1 \text{ torr}) = 23.3 \text{ torr}$$

According to Dalton's law the total vapour pressure, p_{Total}, is the sum of the two partial vapour pressures:

$$p_{\text{Total}} = p_{\text{Benzene}} + p_{\text{Methylbenzene}} = (18.9 \text{ torr}) + (23.3 \text{ torr}) = 42.2 \text{ torr}$$

The total vapour pressure of the mixture is 42.2 torr.

→ To avoid rounding errors it is best to leave the numerical calculation to the final stage.

> **Question 4.14**
>
> (a) The vapour pressures of benzene and methylbenzene are 94.6 and 29.1 torr respectively at 25 °C. What is the total vapour pressure of a mixture that contains six moles of benzene and two moles of methylbenzene?
>
> (b) The vapour pressures of compound A and compound B are 70.2 and 81.3 kPa respectively at room temperature. What is the total vapour pressure of an ideal mixture that contains six moles of compound A and four moles of compound B?

Worked example 4.7D

The vapour pressures of benzene and methylbenzene are 94.6 and 29.1 torr respectively at 25 °C. A mixture contains two moles of benzene and eight moles of methylbenzene. What is the composition of the vapour?

Solution

Benzene and methylbenzene are similar in molecular structure and have similar chemical interactions; the mixture will therefore behave ideally. Use the data provided to determine the mole fraction of benzene in the vapour and hence the composition.

$$x_{\text{Benzene, vapour}} = \frac{p_{\text{Benzene}}}{p_{\text{Total}}} = \frac{p_{\text{Benzene}}}{p_{\text{Benzene}} + p_{\text{Methylbenzene}}}$$

$$x_{\text{Benzene, vapour}} = \frac{x_{\text{Benzene, liquid}} \times p_{\text{Benzene, pure}}}{(x_{\text{Benzene, liquid}} \times p_{\text{Benzene, pure}}) + (x_{\text{Methylbenzene, liquid}} \times p_{\text{Methylbenzene, pure}})}$$

$$x_{\text{Benzene, vapour}} = \frac{0.2 \times (94.6 \text{ torr})}{(0.2 \times 94.6 \text{ torr}) + (0.8 \times 29.1 \text{ torr})} = 0.45$$

→ To avoid rounding errors it is best to leave the numerical calculation to the final stage.

The mole fraction of benzene in the vapour is 0.45 and the mole fraction of methylbenzene in the vapour is 0.55.

> Does the answer make sense? Yes: as expected the vapour is richer in benzene, the more volatile component. The mole fraction of benzene in the vapour is 0.45, whereas the mole fraction of benzene in the liquid is 0.2.

> **Question 4.15**
>
> (a) The vapour pressures of benzene and methylbenzene are 0.040 bar and 0.025 bar respectively at room temperature. A mixture contains one mole of benzene and six moles of methylbenzene. What is the composition of the vapour?
>
> (b) The vapour pressures of pure liquid A and pure liquid B are 535 torr and 471 torr respectively at 30 °C. An ideal mixture contains three moles of liquid A and seven moles of liquid B. What is the composition of the vapour?

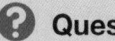

4.8 Two-component mixtures—ideal dilute solutions

As previously discussed, an ideal solution is one where both the solvent and the solute obey Raoult's law. However, many systems are not ideal and deviations away from Raoult's law occur.

An ideal dilute solution is comprised of excess solvent and low concentrations of solute. Raoult's law can only be applied to the solvent. Whilst the vapour pressure of the solute does not follow Raoult's law it does still vary linearly with mole fraction; the proportionality constant is not the vapour pressure of the pure solute but a different constant called Henry's constant. This linear relationship is referred to as Henry's law.

Consider a mixture comprising a solvent A and a solute B. The solvent, A, follows Raoult's law:

$$p_A = x_{A,\,liquid} \times p_{A,\,pure}$$

where:

- p_A is the vapour pressure of component A (units of pressure).
- $x_{A,\,liquid}$ is the mole fraction of A in the liquid mixture (dimensionless).
- $p_{A,\,pure}$ is the vapour pressure of pure A (units of pressure).

Whereas the solute, B, follows Henry's law:

$$p_B = x_{B,\,liquid} \times K_B$$

where:

- p_B is the vapour pressure of component B (measured in units of pressure).
- x_B is the mole fraction of B in the liquid mixture (dimensionless).
- K_B is Henry's law constant (units of pressure).

> Note that in practice Henry's law constant can have a range of units.

Figure 4.8 illustrates how a single component can follow both Henry's law and Raoult's law.

Worked example 4.8A

Carbon dioxide makes up a small proportion of the Earth's atmosphere (~0.04%). Henry's law constant for carbon dioxide is 0.2×10^4 bar. Estimate the normal concentration of carbon dioxide in water.

Solution

First calculate the mole fraction of carbon dioxide in the atmosphere:

0.04% is the same as a mole fraction of 0.0004.

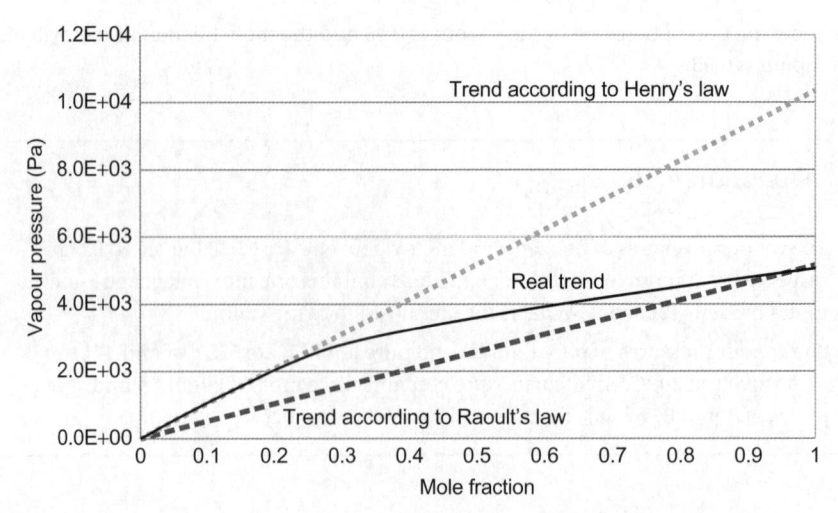

Figure 4.8 Vapour pressure–composition curve illustrating how at low solute mole fraction the system follows Henry's law while at high mole fractions the system follows Raoult's law.

Then calculate the partial pressure of carbon dioxide in the atmosphere:

$$p_A = x_A \times P_{total}$$

$$p_{CO_2, gas} = x_{CO_2} \times P_{total}$$

$$p_{CO_2, gas} = 0.0004 \times 1.013 \text{ bar} = 4 \times 10^{-4} \text{ bar}$$

Use Henry's law to calculate the mole fraction of carbon dioxide in solution:

$$p_B = x_{B, liquid} \times K_B$$

$$p_{CO_2, gas} = x_{CO_2, liquid} \times K_{CO_2}$$

Rearrange to make the mole fraction the subject:

$$x_{CO_2, liquid} = \frac{p_{CO_2, gas}}{K_{CO_2}}$$

$$x_{CO_2, liquid} = \frac{4 \times 10^{-4} \text{ bar}}{0.2 \times 10^4 \text{ bar}} = 2 \times 10^{-7}$$

Next determine the concentration of carbon dioxide in solution in mol dm^{-3} using:

$$x_{CO_2, liquid} = \frac{n_{CO_2}}{n_{CO_2} + n_{H_2O}}$$

The mole fraction of carbon dioxide is very small. Therefore n_{CO_2} is very small and the denominator approximates to n_{H_2O} and the expression becomes:

$$x_{CO_2, liquid} = \frac{n_{CO_2}}{n_{H_2O}}$$

$$2 \times 10^{-7} = \frac{n_{CO_2}}{n_{H_2O}}$$

1 dm^3 of water has a mass of 1000 g, and since the molar mass of water is 18.01 g mol^{-1} then 55.6 mol of water are present.

$$2 \times 10^{-7} = \frac{n_{CO_2}}{55.6 \text{ mol}}$$

$$2 \times 10^{-7} \times 55.6 \text{ mol} = n_{CO_2}$$

$$1 \times 10^{-5} \text{ mol} = n_{CO_2}$$

The normal concentration of dissolved carbon dioxide in water is 1×10^{-5} mol dm^{-3}.

> **Question 4.16**
>
> Nitrogen makes up a large proportion of the Earth's atmosphere (~78%). Given that Henry's law constant for nitrogen is 9×10^4 bar, estimate the normal concentration of nitrogen in water.

4.9 Two-component mixtures—non-ideal binary liquid mixtures

Similar liquid mixtures such as hexane and heptane or benzene and methylbenzene tend to behave in an ideal fashion. However, in many cases the vapour pressure above the solution is either greater than expected or smaller than expected.

Non-ideal behaviour arises if the molecules in solution strongly interact with one another. Positive deviations from Raoult's law occur when the molecular interactions in the solution are weaker than those present in the pure liquids. If the molecular interactions in the solution are weaker the molecules find it easier to enter the vapour phase and as a result the vapour pressure above the solution is higher than that predicted by Raoult's law. An example of a system that exhibits positive deviation is propanone and carbon disulfide.

In contrast, when the molecular interactions in the solution are stronger than those in the pure liquids then negative deviations from Raoult's law occur. If the molecular interactions in the solution are stronger then the molecules find it harder to enter the vapour phase and as a result the vapour pressure above the solution is lower than that predicted by Raoult's law. An example of a system that exhibits negative deviation is propanone and trichloromethane.

Ideal solutions follow Raoult's law, which can be expressed as:

$$p_{A,\,ideal} = x_{A,\,liquid} \times p_{A,\,pure}$$

However, Raoult's law must be modified for real solutions by adding a term known as the activity coefficient, γ:

$$p_{A,\,real} = \gamma_A \times x_{A,\,liquid} \times p_{A,\,pure}$$

where:

- $p_{A,\,real}$ is the real vapour pressure of component A (units of pressure).
- γ_A is the activity coefficient of component A (dimensionless).
- $x_{A,\,liquid}$ is the mole fraction of A in the liquid mixture (dimensionless).
- $p_{A,\,pure}$ is the vapour pressure of pure A (units of pressure).

This expression can be rearranged to make γ_A the subject:

$$\frac{p_{A,\,real}}{x_{A,\,liquid} \times p_{A,\,pure}} = \gamma_A$$

As the denominator of the left-hand side is equivalent to $p_{A,\,ideal}$ it can be seen that the activity coefficient is the ratio of the real vapour pressure to the ideal vapour pressure:

$$\frac{p_{A,\,real}}{p_{A,\,ideal}} = \gamma_A$$

Thus if the activity coefficient is greater than one, positive deviations occur; if it is less than one, negative deviations occur.

Worked example 4.9A

Consider a non-ideal binary solution comprised of 0.32 mol of liquid A and 0.74 mol of liquid B. The saturated vapour pressures of A and B are 49.8 kPa and 31.2 kPa respectively. At equilibrium the vapour pressures of A and B were 39.2 kPa and 34.1 kPa respectively. Calculate the activity coefficients for A and B.

Solution

First determine the mole fraction of A and B:

$$x_{A,\,liquid} = \frac{n_A}{n_A + n_B}$$

$$x_{A,\,liquid} = \frac{0.32}{0.32 + 0.74} = 0.30$$

$$x_{B,\,liquid} = \frac{n_B}{n_A + n_B}$$

$$x_{B,\,liquid} = \frac{0.74}{0.32 + 0.74} = 0.70$$

Then calculate the activity coefficient of A using:

$$\gamma_A = \frac{p_{A,\,real}}{x_{A,\,liquid} \times p_{A,\,pure}}$$

$$\gamma_A = \frac{39.2 \times 10^3 \text{ Pa}}{0.30 \times (49.8 \times 10^3 \text{ Pa})}$$

$$\gamma_A = 2.6$$

Calculate the activity coefficient of B using:

$$\gamma_B = \frac{p_{B,\,real}}{x_{B,\,liquid} \times p_{B,\,pure}}$$

$$\gamma_B = \frac{34.1 \times 10^3 \text{ Pa}}{0.70 \times (31.2 \times 10^3 \text{ Pa})}$$

$$\gamma_B = 1.6$$

The activity coefficients for A and B are 2.62 and 1.56 respectively. These values are larger than 1, indicating that the system is an example of a positive deviation away from Raoult's law—that is, the molecular interactions in the solution are weaker than in the pure components and as a result molecules find it easier to escape into the vapour.

Question 4.17

Consider a non-ideal binary solution comprising 0.650 mol of liquid A and 0.350 mol of liquid B. The saturated vapour pressures of A and B are 41.2 kPa and 31.9 kPa respectively. At equilibrium the vapour pressures of A and B were 36.1 kPa and 28.9 kPa respectively. Calculate the activity coefficients for A and B.

4.10 Two-component mixtures—distillation of binary liquid mixture

When a binary liquid mixture is heated, the vapour in equilibrium with the liquid phase will become richer in the more volatile component. The system is usually represented as a temperature–composition diagram measured at a fixed pressure, typically 1 atm. A temperature–composition diagram is thus a plot of the boiling temperature of the liquid mixture as a function of the composition of the solution (typically mole fraction). It is usual to plot the x-axis with the mole fraction of the more volatile component (or the one with the lower boiling point) increasing.

Worked example 4.10A

An example of a temperature–composition diagram is shown in Figure 4.9 for a binary solution comprised of A and B.

A has a boiling point of 354 K and B a boiling point of 390 K. Thus A is the more volatile component. The lower curve indicates the boiling point of the binary mixture and illustrates how it decreases from the value corresponding to the boiling point of pure B to a value corresponding to the boiling point of pure A as the composition of the mixture changes. The upper curve indicates the vapour composition and shows that for a given liquid composition the vapour is richer in the more volatile component.

Now consider the diagram in more detail. The plot comprises three zones (one zone towards the bottom left, one zone towards the top right and one zone found towards the centre). In the bottom left zone, at temperatures below the mixture boiling point, a single two-component liquid phase exists. In the top right zone, at temperatures above the boiling point, a single two-component vapour phase exists and in the central zone liquid and vapour coexist.

First consider the single phase zones (bottom left and top right). In the two-component liquid phase zone (bottom left) the composition of the phase is that of the mixture overall. Remember the phase rule discussed earlier:

$$F = C - P + 2$$

where F is the number of degrees of freedom (number of intensive variables such as pressure or temperature that can be changed without disturbing the number of phases in equilibrium), C is the number of components in the system, and P is the number of phases.

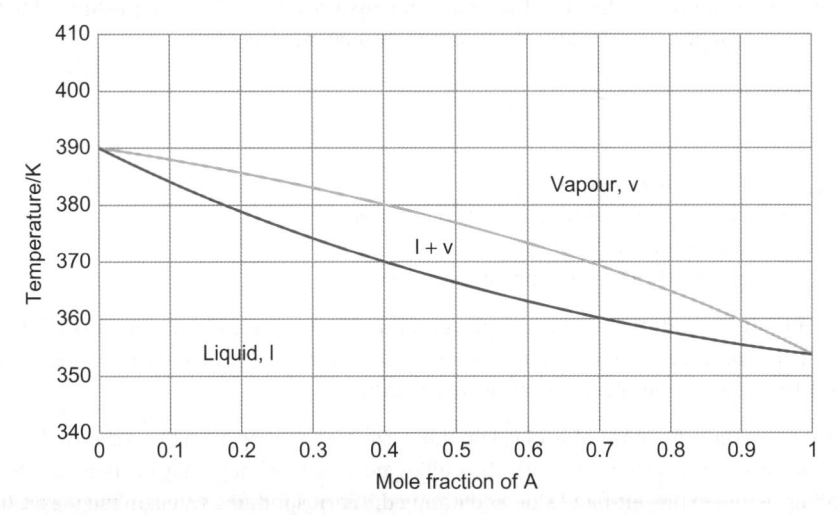

Figure 4.9 Temperature–composition diagram for binary mixture of A and B.

4 PHASE EQUILIBRIUM

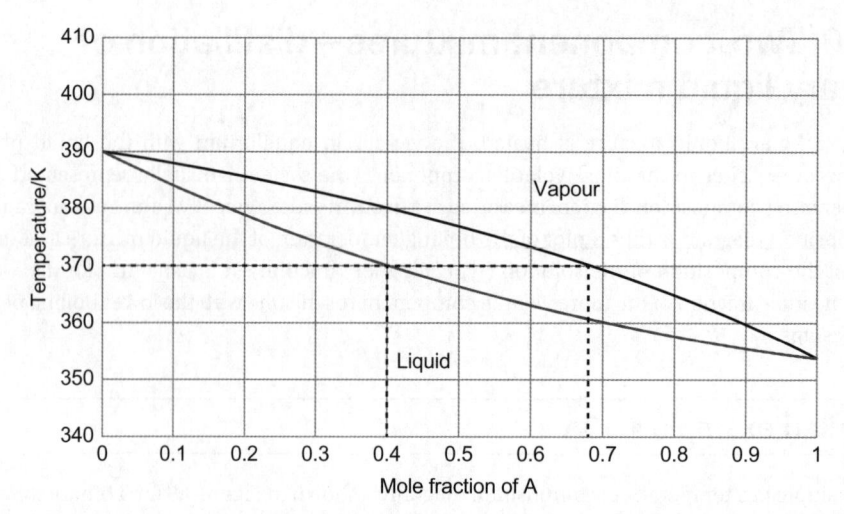

Figure 4.10 Temperature–composition diagram for a binary mixture of A and B, illustrating the effect of heating a mixture where the mole fraction of A is 0.4.

Here:

$$F = 2 - 1 + 2$$
$$F = 3$$

The three degrees of freedom are pressure (in this case fixed at 1 atm), temperature, and mole fraction. (This means that a range of liquid compositions exists over a range of temperatures and it can be seen that the boiling point of this liquid phase depends upon its composition.) Figure 4.10 shows that a mixture where the mole fraction of A is 0.4 has a boiling point of 370 K. Similarly if the temperature is high enough a single two-component vapour phase forms (top right zone) where the composition of the phase is that of the mixture overall.

Next consider the central zone where two phases (the liquid and the vapour) are present. Here there is a two-component system with two phases. If we now apply the phase rule to this two-component, two-phase system:

$$F = 2 - 2 + 2$$
$$F = 2$$

So here there are only two degrees of freedom, the pressure (in this case fixed at 1 atm pressure) and temperature, i.e. the mole fraction of the mixture is fixed. The composition of the liquid and vapour phase are fixed and connected through the lever rule:

$$n_{Liquid} L_{Liquid} = n_{Vapour} L_{Vapour}$$

where:

- n_{Liquid} is the amount in moles of liquid.
- L_{Liquid} is the length from the liquid phase boundary.
- n_{Vapour} is the amount in moles of gas.
- L_{Vapour} is the length from the gas phase boundary.

To interpret a temperature-composition diagram, start by noting the composition of the liquid mixture to be considered. As an example, consider a mixture where the mole fraction of A is 0.4; Figure 4.10 illustrates the effect of heating this mixture.

- Identify the mole fraction value of 0.4 on the x-axis and read up from the x-axis to the liquid composition line (the lower curve). The boiling point of this liquid may be identified by reading across to the left-hand side and noting the corresponding value on the y-axis. In this case it is 370 K.

- The composition of the vapour produced from this mixture may then be identified by reading across to the right-hand side until the vapour composition line is reached (the upper curve) and then reading down to the x-axis to note the mole fraction. In this case the vapour will comprise of A (mole fraction 0.68) and B (mole fraction 0.32).

Note that the vapour is richer in the more volatile component A. If this vapour is condensed and the process repeated then the liquid becomes richer and richer in the more volatile component. A practical application of this theory is the technique of distillation, which involves heating a liquid mixture so that evaporation occurs. The distillate is captured and condensed and is seen to contain a greater proportion of the more volatile component. This process provides a means of separation or purification.

Note on azeotropes: Most solutions are not ideal however and both negative and positive deviations from Raoult's law occur. Negative deviations from Raoult's law result in the formation of a mixture that has a higher boiling point than either of the two components. This means it is not possible to entirely separate the two components by distillation. Instead, a mixture of fixed boiling point, known as an azeotrope, is produced.

Similarly positive deviations from Raoult's law result in the formation of a mixture that has a lower boiling point than either of the two pure components. Again, this means that it is not possible to entirely separate the two components by distillation and again an azeotrope is produced. When an azeotrope has formed no further separation of the components by distillation is possible since the liquid and the vapour have the same composition.

Note on binary liquid mixtures—partial miscibility: Some liquids only mix in other liquids under certain conditions and this can lead to phase separation of the two components. For a mixture with a set composition, the two liquids mix to produce a single liquid phase if the temperature is high enough. If the mixture is then cooled to a sufficiently low temperature phase separation occurs. The composition of the two phases in equilibrium at a given temperature can be obtained from a temperature-composition diagram. The relative amounts of the two phases are given by the lever rule.

Turn to the Synoptic questions section on page 173 to attempt questions that encourage you to draw on concepts and problem-solving strategies from several topics within a given chapter to come to a final answer.

Final answers to numerical questions appear at the end of the book, and full worked solutions appear on the book's website, where you can also find a set of bonus questions for each chapter. Go to www.oxfordtextbooks.co.uk/orc/chemworkbooks/.

5
Reaction kinetics

5.1 The rate of a chemical reaction

The rate of a chemical reaction is the rate at which the reactants are converted into products. We generally monitor this by measuring the change of concentration of a reactant or product with time. This is defined as the **rate of a reaction** and has units of the type mol dm^{-3} s^{-1}. Because the rate of appearance of products and disappearance of reactants changes during the course of the reaction chemists are generally interested in the rate of reaction at a specific time during the reaction. This is called the **instantaneous** rate.

If we measure how long it takes for one of the reactants to be completely converted into products we can obtain the **average rate of reaction**. For a simple chemical reaction, where one mole of reactant A is converted into one mole of product B:

$$A \rightarrow B$$

the average rate of reaction would be given by:

$$\text{Rate} = \text{Rate}_1 = -\frac{\Delta[A]}{\Delta t}$$

$\Delta[A]$ is the change in concentration of A and Δt is the length of time taken. The symbol Δ represents a large change.

You will notice that the equation has a minus sign in front of the expression. This is because reaction rates must always be positive values. The concentration of A is decreasing with time and so $\Delta[A]$ is a negative value. The minus sign must be included to make the rate into a **positive** value.

It might be more convenient to measure the rate of reaction by measuring the change in concentration of B. For example, if B is coloured we can use a colorimeter, or if B is a gas we can measure the increase in pressure. In this case the average rate of reaction would be given by:

$$\text{Rate} = \text{Rate}_2 = \frac{\Delta[B]}{\Delta t}$$

Notice that there is no minus sign in this expression. This is because the concentration of B is increasing and so is a positive value.

Because the rate at which A is used up must be the same as the rate at which B is formed we can say that the average rate of reaction is given by:

$$\text{Rate} = \text{Rate}_1 = \text{Rate}_2$$

If [A] is measured in mol dm^{-3} the units of rate of change of concentration with time will be mol dm^{-3} s^{-1}, assuming the time is measured in seconds.

> If [A]$_0$ is the concentration of the reactant A at the start of the reaction when $t = 0$ and [A]$_f$ is the concentration of A at the end of the reaction, then [A]$_f$ is less than [A]$_0$ and so [A]$_f$ −[A]$_0$ is negative.

> All the expressions of rate must be positive.

Worked example 5.1A

a) Give two equations for the average rate of the following reaction, over the time period, t, in terms of the concentrations of reactants and products:

$$N_2O_4(g) \rightarrow 2\,NO_2(g)$$

b) Show the relationship between the two rates of reaction.

Solution

a) $\text{Rate}_1 = -\dfrac{\Delta[N_2O_4]}{\Delta t}$ for the consumption of N_2O_4.

$\text{Rate}_2 = \dfrac{\Delta[NO_2]}{\Delta t}$ for the formation of NO_2.

→ As $[N_2O_4]$ is decreasing we must include a minus sign to make Rate_1 positive.

b) Because one mole of N_2O_4 produces two moles of NO_2, NO_2 is being formed at twice the rate that N_2O_4 is being used up, so $\text{Rate}_2 = 2 \times \text{Rate}_1$ and $\text{Rate}_1 = \frac{1}{2}\,\text{Rate}_2$ and

$$-\dfrac{\Delta[N_2O_4]}{\Delta t} = \dfrac{1}{2}\dfrac{\Delta[NO_2]}{\Delta t}$$

We say that the average rate of reaction with respect to NO_2 is twice the rate of reaction with respect to N_2O_4.

> **? Question 5.1**
>
> (a) Give an expression for the average rate of the following reaction (Rate_1) in terms of the rate of formation of $NH_3(g)$ with time, t.
>
> $$3\,H_2(g) + N_2(g) \rightarrow 2\,NH_3(g)$$
>
> (b) Give a further expression for the average rate of the reaction (Rate_2) in terms of the rate of disappearance of H_2 with time, t.
>
> (c) What is the relationship between Rate_1 and Rate_2?

The rates of most chemical reactions change as reaction proceeds. Generally the rate is quite fast at the start of the reaction when there are plenty of reactants present. However, as the reactants are used up the rate will slow down. Chemists are generally interested in the rate of reaction at a specific time, t. This is known as the instantaneous rate. If we plot the concentration of a reactant against the time taken for the reactant to be used up a curve such as in Figure 5.1 may be obtained.

→ The **instantaneous rate** is the rate of reaction at a specific time during the reaction.

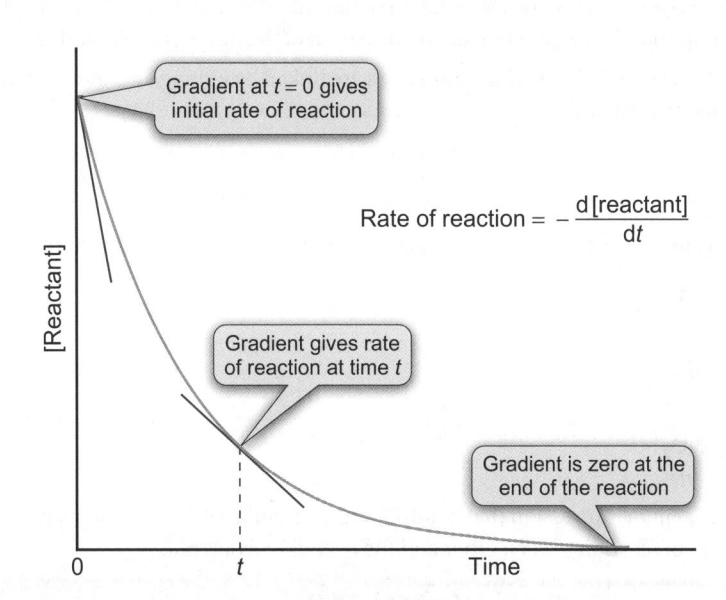

Figure 5.1 Plot of concentration against time for a reactant.
Reproduced from Burrows et al, *Chemistry*[3] second edition (Oxford University Press, 2013). © Andrew Burrows, John Holman, Andrew Parsons, Gwen Pilling, and Gareth Price 2013

The rate of reaction at any point during the reaction is given by the gradient of the tangent to the curve at that point. Mathematically this is equivalent to saying the rate at that point is the small change in concentration of the reactant divided by the small change in time. The gradient of the tangent is negative because the concentration of the reactant is decreasing. Consequently, we have to include a minus sign to make the rate into a positive value. So:

$$\text{Rate of reaction} = -\frac{d[\text{reactant}]}{dt}$$

> The small letter 'd' represents an infinitessimally small change.

In the same way we can plot the change in concentration of the product with time and obtain the rate of reaction at any point, t, by again measuring the gradient of the tangent to the curve at this point. As the product is increasing in concentration, the gradient is positive and so the rate of reaction at any time can be expressed by:

$$\text{Rate of reaction} = \frac{d[\text{product}]}{dt}$$

> A tangent to a curve is a straight line which just touches the curve at a given point. The gradient of the tangent is the same as the gradient of the line at that point. The gradient of the tangent is found in the same way as the gradient of a straight line by using the equation:
> gradient = $\frac{y_2 - y_1}{x_2 - x_1}$ where (x_1, y_1), (x_2, y_2) are points as shown on the plot.

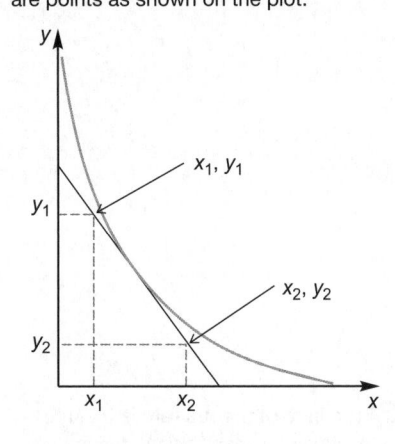

At the beginning of the reaction when time $t = 0$ we have the initial rate of reaction. At the end of the reaction, or when the reaction has reached equilibrium and there is no further change in concentration of reactants or products, the rate of the forward and backward reactions will be the same.

For a general reaction such as:

$$a\,A + b\,B \rightarrow c\,C + d\,D$$

the rate of reaction can be expressed either in terms of disappearance of reactants or appearance of products. These expressions are linked in the following way in the definition for the rate of the reaction:

$$\text{rate of reaction} = -\frac{1}{a}\frac{d[A]}{dt} = -\frac{1}{b}\frac{d[B]}{dt} = \frac{1}{c}\frac{d[C]}{dt} = \frac{1}{d}\frac{d[D]}{dt}$$

Worked example 5.1B

Dinitrogen pentoxide dissociates according to the following chemical equation.

$$2\,N_2O_5(g) \rightarrow 4\,NO_2(g) + O_2(g)$$

(a) Use differential expressions to write three rate equations in terms of the rate of consumption of N_2O_5 and the rate of formation of the products, NO_2 and O_2.

(b) Show how these individual reaction rates are linked by writing an expression for the overall rate of reaction.

Solution

(a) Write expressions for the rates of reaction in terms of each of the reactants and products:

$$\text{Rate}_1 = -\frac{d[N_2O_5]}{dt}$$

$$\text{Rate}_2 = \frac{d[NO_2]}{dt}$$

$$\text{Rate}_3 = \frac{d[O_2]}{dt}$$

(b) Use the general expression that combines the stoichiometric coefficients to relate each of the rates of reaction to each other and the overall reaction rate.

$$\text{Rate of reaction} = -\frac{1}{2}\frac{d[N_2O_5]}{dt} = \frac{1}{4}\frac{d[NO_2]}{dt} = \frac{d[O_2]}{dt}$$

> Check your answer makes sense: NO_2 is being produced at the fastest rate so we have to divide by the largest number (4) to obtain equivalent reaction rates. N_2O_5 is the only reactant being consumed so that rate with respect to N_2O_5 is negative.

Worked example 5.1C

Methane undergoes combustion according to the following reaction:

$$CH_4(g) + 2 O_2(g) \rightarrow CO_2(g) + 2 H_2O(g)$$

(a) At a certain time the rate of consumption of oxygen was found to be 0.2 mol dm^{-3} s^{-1}.

   i. What is the rate of consumption of methane at this time?
   ii. What is the rate of production of gaseous CO$_2$ at this time?

(b) If initially there were 10.0 mol of CH$_4$ and the average rate of consumption of oxygen was 0.1 mol dm^{-3} s^{-1}, how many moles of CH$_4$ would be present after 20 s?

Solution

a) i. The rate of consumption of methane at any time is half the rate of consumption of oxygen. So at this time the rate of consumption of methane will be 0.2/2 mol dm^{-3} s^{-1} which gives a value of 0.1 mol dm^{-3} s^{-1}.

   ii. The rate of production of CO$_2$ is the same as the rate of consumption of CH$_4$ at that time, so the answer will also be 0.1 mol dm^{-3} s^{-1}.

b) If the average rate of consumption of oxygen during the reaction is 0.1 mol dm^{-3} s^{-1}, then the average rate of consumption of methane during this time will be 0.05 mol dm^{-3} s^{-1}. During 20 s the amount of methane consumed will be 20 × 0.05 = 1.0 mol. The amount of methane remaining is therefore (10.0 − 1.0) = 9.0 mol.

Question 5.2

The persulfate ion (S$_2$O$_8^{2-}$) reacts in aqueous solution with iodide ions to produce the triiodide ion (I$_3^-$). The reaction can be followed by the rate of appearance of the triiodide ion as this is the only species that absorbs visible light.

$$S_2O_8^{2-}(aq) + 3 I^-(aq) \rightarrow 2 SO_4^{2-}(aq) + I_3^-(aq)$$

(a) State which of the reactants and products is changing concentration at the fastest rate.
(b) Write expressions for the rate of disappearance of each reactant and the rate of appearance of each product with time.
(c) Use the general expression which combines the stoichiometric coefficients to relate each of the rates of reaction to each other and the overall reaction rate.

Question 5.3

In the following reaction:

$$H_2SeO_3(aq) + 6 I^-(aq) + 4 H^+(aq) \rightarrow Se(s) + 2 I_3^-(aq) + 3 H_2O(l)$$

The rate of formation of triiodide ion was found to be 1×10^{-5} mol dm^{-3} s^{-1}. What is the rate at which the iodide ions are being consumed in the reaction?

> **? Question 5.4**
>
> Bromate ions (BrO_3^-) react with bromide ions (Br^-) in acidic solution to produce bromine (Br_2) according to the following equation.
>
> $$5\,Br^-(aq) + BrO_3^-(aq) + 6\,H^+ \rightarrow 3\,Br_2(aq) + 3\,H_2O(l)$$
>
> The rate of reaction can be studied colorimetrically by the rate of production of bromine, which was found to be 0.12 mol s^{-1}. If initially there were 0.05 mol Br^- (aq) in the solution how many moles of Br^- would be left after 0.1 seconds?

5.2 The order of a chemical reaction

For any chemical reaction such as:

$$a\,A + b\,B \rightarrow c\,C + d\,D$$

The rate equation can be written as:

$$\text{Rate of reaction} = k\,[A]^m[B]^n$$

where:

- m is the order of reaction with respect to A.
- n is the order of reaction with respect to B.
- k is the **rate constant**.

→ Note that there is no relationship between the order of reaction and the stoichiometry of the chemical equation. The rate equation cannot be determined from the stoichiometric equation but must be determined experimentally.

The overall order of the reaction is given by $m + n$.

The orders of reactions with respect to individual reactants are typically, but not always, small whole numbers such as 0, 1, 2.

Only if the reaction $a\,A + b\,B \rightarrow$ **products** is a single step reaction that goes to completion can the rate of reaction be written as: Rate of reaction = $k\,[A]^a[B]^b$.

→ A single step reaction is one that takes place in only one step as the name suggests. Most chemical reactions take place via a series of steps known as the reaction mechanism. If there is only one step in the reaction the rate of reaction must depend upon the rate of this step only.

Worked example 5.2A

The rate of the reaction:

$$2\,N_2O_5 \rightarrow 4\,NO_2 + O_2$$

is governed by the rate equation: Rate = $k\,[N_2O_5]$

(a) What is the order of reaction with respect to:
   i. N_2O_5;
   ii. NO_2;
   iii. O_2?

(b) What is the overall order of the reaction?

Solution

→ Any number raised to the power 0 = 1, so $10^0 = 1$. Any number raised to the power 1 is the same as the number, so $10^1 = 10$. Any number raised to the power 2 means the number is squared (or multiplied by itself). So $10^2 = 100$ (the same as 10×10).

(a) i) The order with respect to N_2O_5 is the power that N_2O_5 is raised to in the rate equation. In this case the power is 1, as any number to the power 1 is the number itself.

ii) and iii) Neither NO_2 nor O_2 appear in the rate equation and so the rate of the reaction at any time cannot depend on the concentration of either of these two species. Formally the order of reaction with respect to NO_2 and O_2 is therefore zero.

(b) The overall order of the reaction is the sum of the individual orders, so this must be equal to 1 in this case. If the overall order is 1 we say it is a first-order reaction.

Worked example 5.2B

The experimentally determined rate equation for the reaction:

$$2\,NO\,(g) + O_2\,(g) \rightarrow 2\,NO_2\,(g)$$

is:

rate of reaction, $Rate_1 = k[NO]^2[O_2]$.

If the concentrations of both starting materials are doubled, by how much will the initial rate of reaction increase?

Solution

If the concentrations of both NO_2 and O_2 are doubled then the new rate of reaction, $Rate_2$, will be:

$Rate_2 = k(2 \times [NO])^2 (2 \times [O_2])$.

Multiplying out terms, $Rate_2$ becomes:

$Rate_2 = k \times 4 \times [NO]^2 \times 2 \times [O_2] = 8k[NO]^2[O_2]$.

This expression is eight times larger than the original expression for the rate of reaction, $Rate_1 = k[NO]^2[O_2]$, and so we can say the initial rate increases by a factor of eight.

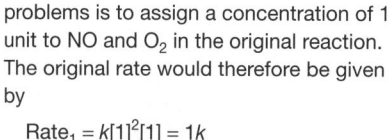

→ Another way to solve these types of problems is to assign a concentration of 1 unit to NO and O_2 in the original reaction. The original rate would therefore be given by

$Rate_1 = k[1]^2[1] = 1k$

If the new concentrations are twice the original ones, then they are now equivalent to 2 units so the new rate, $Rate_2$, becomes $k[2]^2[2] = 8k$. This value is eight times the original one therefore the rate increases by a factor of eight.

❓ Question 5.5

The rate of the following reaction:

$$2\,NO\,(g) + 2\,H_2\,(g) \rightarrow N_2\,(g) + 2\,H_2O\,(g)$$

can be expressed by the rate equation:

$$-\frac{d[NO]}{dt} = k[NO]^2[H_2]$$

(a) What is the order of the reaction with respect to NO and H_2 and hence the overall order of the reaction?

(b) Write an expression for the rate of reaction with respect to the rate of formation of nitrogen gas. How is this rate expression related to the rate of disappearance of nitric oxide?

(c) If the concentration of NO is increased by five but the concentration of hydrogen remains constant by what factor will the rate increase?

❓ Question 5.6

Experiments have shown that the substitution reaction of chloromethane, CH_3Cl, by hydroxide ions is an S_N2 reaction which means it proceeds by nucleophilic substitution where both reactants are involved in the rate-limiting step. This type of reaction is known as a bimolecular reaction.

(a) Suggest an expression for the rate equation for this reaction.

(b) What is the order of the reaction with respect to:
  i. CH_3Cl;
  ii. OH^-?

(c) What is the overall order of the reaction?

> **Question 5.7**
>
> The rate equation for the reaction:
>
> $NO_2(g) + CO(g) \rightarrow NO(g) + CO_2(g)$
>
> is given by:
>
> Rate of reaction = $k[NO_2]^2$
>
> where k is the rate constant for the reaction.
>
> If the concentration of NO_2 increases by a factor of three, by how much will the rate of reaction change?

5.3 Initial rates method for determining the rate equation

→ Experimentally there are different ways of processing and interpreting data from the initial rates method. You should refer to your textbook for more details.

The initial rates method for determining rate equations takes a series of reactions with different initial concentrations of a reactant. The initial rate of reaction is determined for each run by measuring the tangent to the concentration against time curve at the start of reaction. If the reaction involves several different reactants a series of experiments must be conducted in which only one reactant concentration is varied and the others kept constant. In this way the effect of changing the concentration upon the rate can be determined as in the following example.

Worked example 5.3A

The formation of gaseous hydrogen iodide, HI (g), was monitored at a temperature of 298 K:
$H_2(g) + I_2(g) \rightarrow 2HI(g)$

The following data relating to the rate of reaction for each run were collected:

Experiment number	Initial conc. [H_2]/mol dm^{-3}	Initial conc. [I_2]/mol dm^{-3}	Initial rate of reaction/mol dm^{-3} s^{-1}
1	0.100	0.200	8.0×10^{-3}
2	0.200	0.200	16.0×10^{-3}
3	0.100	0.050	2.0×10^{-3}

(a) Derive an expression for the rate equation.
(b) Calculate k, the rate constant for the reaction at 298 K, stating its units.
(c) Predict the initial rate of reaction at 298 K if the initial concentration of hydrogen is 0.050 mol dm^{-3} and iodine is 0.050 mol dm^{-3}.

Solution

(a) In order to derive an expression for the rate equation we must find out how the rate of reaction depends upon the concentrations of the reactants.

Looking at how the rate depends upon hydrogen, we must first choose two reaction runs where the concentration of hydrogen is changing and iodine is kept constant.

In experiments (1) and (2) where [I_2] is constant, doubling the hydrogen concentration causes the rate of reaction to increase from 8.0×10^{-3} mol dm^{-3} to 16.0×10^{-3} mol dm^{-3}. The rate of reaction is therefore directly proportional to the concentration of hydrogen so the rate is first-order in hydrogen.

We can show this mathematically by writing two separate rate equations for experiments (1) and (2) and letting the order of reaction with respect to hydrogen be n.

Experiment (1): $\text{Rate}_1 = k[H_2]^n$. Substituting values 8.0×10^{-3} mol dm^{-3} = $k[0.1]^n$.

Experiment (2): $\text{Rate}_2 = k[H_2]^n$. Substituting values 16.0×10^{-3} mol dm^{-3} = $k[0.2]^n$.

Dividing Rate$_2$ by Rate$_1$ we obtain:

$$\frac{16.0 \times 10^{-3}}{8.0 \times 10^{-3}} = \frac{k[0.200]^n}{k[0.100]^n}$$

The rate constant is the same for each experiment if the temperature is constant and so cancels.

The equation therefore simplifies to $2 = [2]^n$.

So n, the order of reaction with respect to hydrogen, must be one in this case and so the reaction is first order.

To work out the order of reaction with respect to iodine we have to choose two experiments where the hydrogen concentration is kept constant and the iodine concentration changes. For this we can choose experiments (1) and (3) and let the order of reaction with respect to iodine = m. By inspection we can see that the concentration of iodine in reaction (1) is four times the concentration of iodine in reaction (3). The rate of reaction (1) is also four times the rate of reaction (3). From this we can see that the rate of reaction is directly proportional to the iodine concentration, so the reaction is also first order with respect to iodine.

Again this can be shown mathematically by the following:

Experiment (1): $\text{Rate}_1 = k[I_2]^m$. Substituting values 8.0×10^{-3} mol dm^{-3} = $k[0.200]^m$.

Experiment (3): $\text{Rate}_3 = k[I_2]^m$. Substituting values 2.0×10^{-3} mol dm^{-3} = $k[0.050]^m$.

Dividing the rate equation for experiment (1) by the equation for experiment (3) we obtain:

$$\frac{8.0 \times 10^{-3}}{2.0 \times 10^{-3}} = \frac{k[0.200]^m}{k[0.050]^m}$$

$$4 = [4]^m$$

So again, $m = 1$ and the reaction is first order in iodine also. The overall rate of reaction is therefore:

$\text{Rate} = k[H_2][I_2]$

(b) To calculate the rate constant we need to take one set of data, say that for the first experiment, and insert the values into our rate equation determined in (a) above.

$\text{Rate} = k[H_2][I_2]$

8.0×10^{-3} mol dm^{-3} s^{-1} = k (0.100 mol dm^{-3}) × (0.200 mol dm^{-3})

So $k = \dfrac{8.0 \times 10^{-3} \text{ mol dm}^{-3} \text{ s}^{-1}}{0.020 \text{ mol}^2 \text{ dm}^{-6}} = 0.40$ mol^{-1} dm^3 s^{-1}

(c) To find the rate at any time once we have the rate constant we can insert the concentrations of the reactants in the rate equation:

Rate = 0.40 mol^{-1} dm^3 s^{-1} × 0.050 mol dm^{-3} × 0.050 mol dm^{-3} = 1.0×10^{-3} mol dm^{-3} s^{-1}.

Worked example 5.3B

In a hypothetical reaction X + Y → Products, the dark spheres in Figure 5.2 represent molecules of X and the light spheres represent molecules of Y. The rate of reaction is first order in both X and Y. The reaction is carried out with different starting concentrations of X and Y represented pictorially in boxes A, B, C, and D of Figure 5.2. Which of the starting conditions (A, B, C, or D) will result in the highest initial rate of reaction?

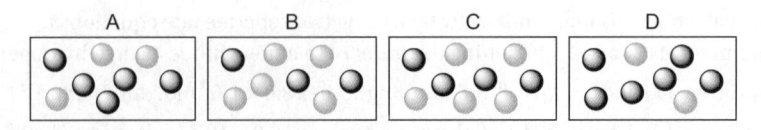

Figure 5.2 Four starting conditions showing differing concentration ratios of X (dark spheres) and Y (light spheres).

Solution

If there are only two reactants, X and Y, and we know the reaction is first order in each, we can write a rate equation:

Rate = k[X][Y]

The dark and light spheres represent molecules of X and Y respectively. The initial rate will therefore be greatest when the product of [X] and [Y] is greatest. Assuming the volume is constant by counting the number of spheres of X and Y in each run the reaction with the greatest value of [X][Y] is reaction C, where we have four molecules of X and five molecules of Y. So:

Rate = [20] × k.

→ We don't know the units here but it doesn't matter as we are simply asked to find which starting conditions will lead to the highest initial rate of reaction.

Question 5.8

Chlorine oxide (ClO·) radicals are removed in the atmosphere by nitrogen dioxide in the presence of dinitrogen, which acts as a type of catalyst as it is regenerated at the end of the reaction.

ClO·(g) + NO$_2$(g) + N$_2$(g) → ClONO$_2$(g) + N$_2$(g)

The reaction was studied by the method of initial rates and the following data measured.

Run	[ClO·(g)]/mol dm^{-3}	[NO$_2$(g)]/mol dm^{-3}	[N$_2$(g)]/mol dm^{-3}	Initial rate of reaction/mol dm^{-3} s^{-1}
1	2.0×10^{-4}	4.0×10^{-4}	6.0×10^{-4}	7.0×10^{-4}
2	1.0×10^{-4}	4.0×10^{-4}	6.0×10^{-4}	3.5×10^{-4}
3	2.0×10^{-4}	8.0×10^{-4}	6.0×10^{-4}	1.41×10^{-3}
4	1.0×10^{-4}	4.0×10^{-4}	1.20×10^{-3}	7.2×10^{-4}

Derive a rate equation for the reaction and calculate a value for the rate constant, k.

Question 5.9

The reaction of HCO$_2$H (aq) and Br$_2$ (aq) is first order with respect to Br$_2$ and zero order with respect to HCO$_2$H. If the concentration of both reactants increases by a factor of three, by what factor will the rate of reaction increase?

Question 5.10

For a hypothetical reaction: X + Y → Products, the dark spheres in Figure 5.3 represent molecules of X and the light spheres represent molecules of Y. The rate of reaction is second order with respect to X but first order with respect to Y. Four experiments are carried out with different starting concentrations represented pictorially in Figure 5.3 in the boxes A, B, C, and D. Which of the starting conditions (A, B, C, or D) will result in the highest initial rate of reaction?

A   B   C

D

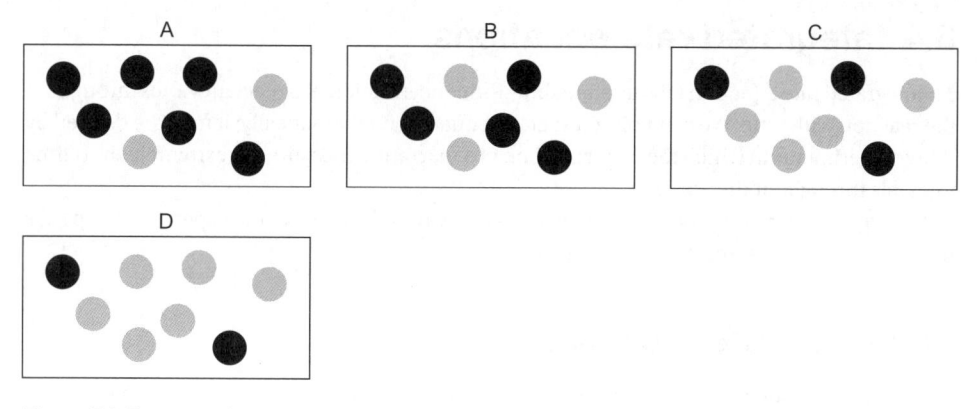

Figure 5.3 Four experimental starting conditions showing differing concentration ratios of X (dark spheres) and Y (light spheres).

❓ Question 5.11

The reduction of nitrogen dioxide by carbon monoxide is represented by the following equation:

$$NO_2(g) + CO(g) \rightarrow NO(g) + CO_2(g)$$

The reaction obeys the rate equation, rate of reaction = $k[NO_2]^2$.

Two experiments were carried out with two different starting conditions:

Experiment	[NO₂]/mol dm⁻³	[CO]/mol dm⁻³
1	0.2	0.2
2	0.4	0.3

(a) What are the relative initial rates of reaction for each experiment?

(b) How would the initial rates of reaction be affected if the concentration of each starting material were decreased by one half?

(c) How would the initial rates of reaction be affected if the volume of the container in each experiment was doubled?

❓ Question 5.12

Hydrogen iodide dissociates at an elevated temperature according to the following equation: $2\,HI(g) \rightarrow H_2(g) + I_2(g)$

The reaction is carried out with different starting concentrations of HI at 435 °C and the data obtained are as follows.

[HI]/mol dm⁻³	0.04	0.02	0.01
Rate/mol dm⁻³ s⁻¹	4.8×10^{-2}	1.2×10^{-2}	3.0×10^{-3}

Suggest a likely rate equation for the reaction based on the data and calculate the rate constant for the reaction.

5.4 Integrated rate equations

An integrated rate equation tells us about how the concentration of a reactant varies throughout the reaction with time. We can't obtain such an equation mathematically; it must be derived by doing experiments in which the concentration of a reactant is measured at extremely short time intervals throughout the reaction.

The way in which the concentration of a reactant varies throughout the experiment depends upon the order of that reactant in the rate equation.

Zero-order rate equations

If the rate of reaction is constant throughout the reaction and doesn't depend upon the concentration of the reactant the reaction is said to be **zero order**.

For a general zero-order reaction: A → products:

$$\text{Rate} = -\frac{d[A]}{dt} = k[A]^0 \quad [A]^0 \text{ means } [A] \text{ raised to the power zero}$$

$[A]^0 = 1$, so $\text{Rate} = -\frac{d[A]}{dt} = k$

A plot of concentration against time gives a straight line with a negative gradient. The gradient of the line is $-k$, the rate constant, as depicted in Figure 5.4a.

A plot of rate against time is therefore a horizontal line parallel to the x axis with a value equal to the rate constant, k, as shown in Figure 5.4b.

The relationship can be obtained by integrating the rate equation.

For a reaction that is zero-order the rate equation is given by:

$$-\frac{d[A]}{dt} = k$$

$$d[A] = -k\,dt$$

$$\int_{[A]_0}^{[A]_t} d[A] = -k\int_0^t dt$$

On integration of the above expression we obtain:

$$[[A]]_0^t = -k[t]_0^t$$

$$([A]_t - [A]_0) = -k(t - 0)$$

which can be rearranged to: $[A]_t = [A]_0 - kt$.

This is known as the integrated rate equation for a zero-order reaction and is the equation of a straight line, $y = c + mx$. It is actually the plot we obtain if the concentration of A is plotted against t for the reaction as in Figure 5.4a. The gradient is $-k$ and the concentration at time, $t = 0$, is the intercept on the y axis.

First-order rate equations

If we have an elementary reaction (one in which there is only one step) whose equation is A → products the variation of [A] with time follows the curve shown in Figure 5.5a.

The reaction would be first-order in [A] and the rate equation given by:

$$\text{Rate of reaction} = -\frac{d[A]}{dt} = k[A]$$

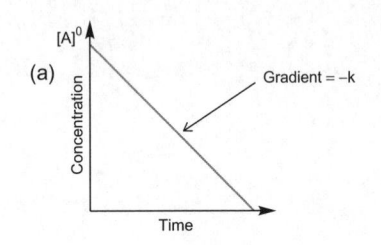

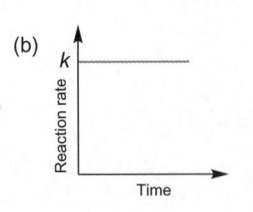

Figure 5.4 (a) The concentration of a reactant with time in a zero-order reaction. (b) The rate of reaction with time in a zero-order reaction.

→ Integration is a mathematical operation. If you haven't covered integration don't worry. The working is shown here and can be found in most textbooks on reaction kinetics, but you wouldn't normally need to derive these equations yourself.

5.4 INTEGRATED RATE EQUATIONS

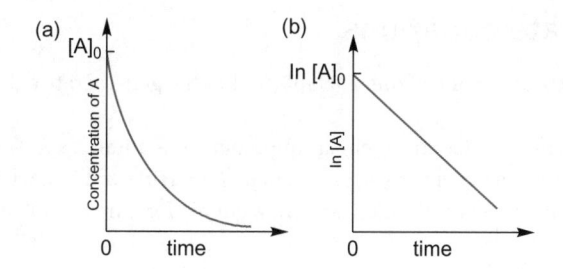

Figure 5.5 (a) A plot of concentration of a reactant (A) against time in a first-order reaction. (b) A plot of ln[A] against time in a first-order reaction. This line has the equation $y = mx + c$.

The integrated rate equation for a first-order reaction is given by:

$$\ln[A]_t = \ln[A]_0 - kt$$

The integrated rate equation shows how the concentration of the reactant at any time t, $[A]_t$, is linked to the initial concentration of the reactant, $[A]_0$ and the rate constant, k.

We call the equation in this form **the integrated rate equation for a first-order reaction**. The equation has the form of a straight line, $y = mx + c$.

The variable y is equivalent to $\ln[A]_t$ and is plotted on the y axis.

The other variable x is equivalent to t and is plotted on the x axis.

If the reaction is first-order in A and we plot $\ln[A]_t$ against t in this way we obtain a straight line with gradient $-k$ and intercept on the y axis of $\ln[A]_0$ as can be seen in Figure 5.5b.

For a first-order reaction A → products we can write the rate equation:

$$\frac{d[A]}{dt} = -k[A]$$

This tells us how the rate of reaction varies with the concentration of the reactant A throughout the progress of the reaction. To find how the actual concentration of A varies throughout the reaction we can integrate both sides of the above equation.

The equation is first rearranged to be written as:

$$\frac{d[A]}{[A]} = -k dt$$

We then integrate both sides of the equation between the conditions at the start of the reaction, time $t = 0$ and any time throughout the reaction, t.

$$\int_{[A]_0}^{[A]_t} \frac{d[A]}{[A]} = -k \int_0^t dt \qquad (5.1)$$

This uses the standard integral $\int_{x_1}^{x_2} \frac{dx}{x} = [\ln x]_{x_1}^{x_2} = \ln x_2 - \ln x_1$

Applying this to equation 5.1 above gives:

$$[\ln[A]]_{[A]_0}^{[A]_t} = -k[t]_0^t$$

$$\ln[A]_t - \ln[A]_0 = -k(t - 0)$$

and we can rearrange this to:

$$\ln[A]_t = \ln[A]_0 - kt$$

You can think of integration as being a mathematical black box if you haven't covered it in maths. You should still be able to apply and use the integrated rate equation even if you don't understand the maths used to derive it.

→ The symbol **ln** means natural logarithm. Natural logarithms are logs to the base e where e = 2.7183. Normal logarithms are to the base 10 and these are represented by **log**.

→ To understand where the integrated rate equation comes from you have to be able to use calculus. It isn't essential to be able to derive the integrated rate equation so a brief explanation is given here of how it is obtained by integrating the rate equation. Again, don't worry if you haven't covered integration.

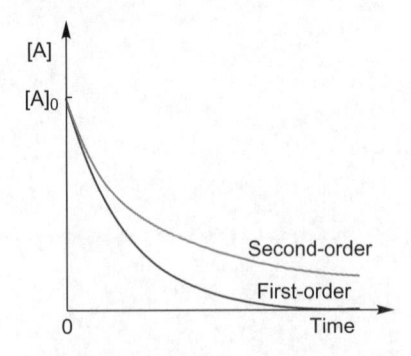

Figure 5.6 A comparison of the concentration against time curves for a first- and a second-order reaction. Reproduced from Burrows et al, *Chemistry*[3] second edition (Oxford University Press, 2013). © Andrew Burrows, John Holman, Andrew Parsons, Gwen Pilling, and Gareth Price 2013.

Second-order rate equations

A reaction that is second order in one of the reactants, A, has the general rate equation: Rate of reaction = $k[A]^2$.

This could be an elementary reaction which has the general equation: A + A → products.

A plot of concentration of A against time is not as steep as for a first-order reaction. Figure 5.6 shows a comparison of the concentration against time curves for a first- and a second-order reaction.

The integrated rate equation for a second-order reaction is given by:

$$\frac{1}{[A]_t} = \frac{1}{[A]_0} + kt$$

Again this equation has the form of a straight line $y = mx + c$ where $\frac{1}{[A]_t}$ is plotted on the y axis and t on the x axis. The gradient in this case is positive and equal to the rate constant, k. The intercept on the y axis gives us the initial concentration of the reactant A as it is equal to $\frac{1}{[A]_0}$.

The equation which allows us to explore second-order kinetics is derived by integrating the rate equation for a second-order reaction. We can write the rate equation as: $-\frac{1}{2}\frac{d[A]}{dt} = k[A]^2$

Rearranging this as: $\frac{d[A]}{[A]^2} = -2k\,dt$ gives us an equation which can be integrated between the limits $t = 0$, when $[A] = [A]_0$, and t when $[A] = [A]_t$.

$$\int_{[A]_0}^{[A]_t} \frac{d[A]}{[A]^2} = -2k \int_0^t dt$$

The standard integral $\int_{[x]_1}^{[x]_2} \frac{dx}{x^2} = \left[-\frac{1}{x}\right]_{x_1}^{x_2} = \left(-\frac{1}{x_2}\right) - \left(-\frac{1}{x_1}\right) = \frac{1}{x_1} - \frac{1}{x_2}$

So integrating and substituting values at the limits for the integral above gives us:

$$\left[-\frac{1}{[A]}\right]_{[A]_0}^{[A]_t} = -2k[t]_0^t = \frac{1}{[A]_0} - \frac{1}{[A]_t} = -(2kt - 2k0) = -2kt$$

Rearranging this becomes:

$$\frac{1}{[A]_t} = \frac{1}{[A]_0} + 2kt$$

The value of k is often combined with the constant to give the following expression:

$$\frac{1}{[A]_t} = \frac{1}{[A]_0} + k't$$

Worked example 5.4A

In a reaction that is zero order in A, the initial concentration of A was 1.2 M and after 120 s the concentration had dropped to 0.40 M. Determine the rate constant for the reaction.

Solution

For a zero-order reaction the rate of reaction can be expressed by $-\frac{d[A]}{dt} = k$ and the integrated rate equation for a zero-order reaction is given by $[A]_t = [A]_0 - kt$.

So rearranging to get an expression for k:

$$k = \frac{[A]_0 - [A]_t}{t} = \frac{(1.2 - 0.40)\,M}{120\,s} = \frac{0.80\,M}{120\,s} = 6.7 \times 10^{-3}\,M\,s^{-1}$$

Worked example 5.4B

The dissociation of sulfuryl chloride, SO_2Cl_2, is first-order with respect to SO_2Cl_2:

$$SO_2Cl_2 (g) \rightarrow SO_2 (g) + Cl_2 (g)$$

The reaction was performed at 298 K and the following data were obtained:

Time/s	[SO_2Cl_2]/mol dm^{-3}
0	1.000
2500	0.947
5000	0.895
7500	0.848
10000	0.803

(a) State the rate equation for the dissociation.
(b) Using a graphical method, calculate the value of the rate constant, k, giving its units.

Solution

(a) Because this is a first-order reaction the rate equation for the dissociation of sulfuryl chloride can be written as:

$$-\frac{d[SO_2Cl_2]}{dt} = k[SO_2Cl_2]$$

(b) As the reaction is first order in SO_2Cl_2 the integrated rate equation can be written as:

$$\ln[SO_2Cl_2]_t = \ln[SO_2Cl_2]_0 - kt$$

This equation has the form of a straight line $y = mx + c$ if we equate $\ln[SO_2Cl_2]_t$ to the variable y and time, t, to the variable x. So the graph required is a plot of $\ln[SO_2Cl_2]_t$ on the y axis and t on the x axis. The table of data can be adapted to include a column for $\ln[SO_2Cl_2]$. Because this is a logarithmic quantity it has no units.

Time/s	[SO_2Cl_2]/mol dm^{-3}	$\ln[SO_2Cl_2]$
0	1.000	0
2500	0.947	−0.054
5000	0.895	−0.111
7500	0.848	−0.165
10000	0.803	−0.219

So the graph plotted should look like the one shown in Figure 5.7.

The equation for the line is $\ln[SO_2Cl_2]_t = \ln[SO_2Cl_2]_0 - kt$ which has the form $y = c + mx$. So the gradient of the line gives us the rate constant, $-k$. From the plot we can see that the line slopes backwards and so the gradient is negative. We therefore expect to get a negative value for the gradient which will result in a positive value for k.

The gradient is given by the change in a value of y divided by the equivalent change in the x value $= \dfrac{y_2 - y_1}{x_2 - x_1}$

This can be done manually if the graph is drawn by hand or can be found from the equation of the line if plotted using Excel. Either way the gradient can be shown to be $-2.0 \times 10^5 \text{ s}^{-1}$. The gradient is equivalent to the negative value of the rate constant, i.e. $-k$. k is therefore $2.0 \times 10^{-5} \text{ s}^{-1}$.

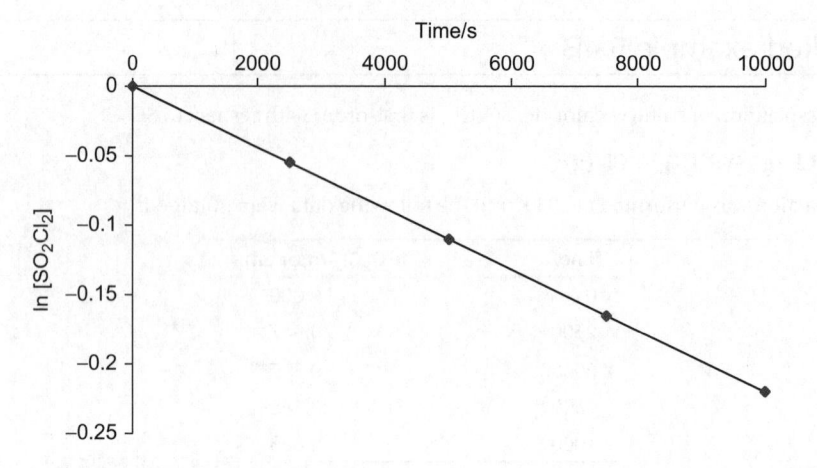

Figure 5.7 Graph of ln[SO$_2$Cl$_2$] against time.

Worked example 5.4C

The decomposition of gaseous nitrogen dioxide, NO$_2$, was monitored at 298 K and the following data were collected. The equation for the decomposition is:

$$NO_2(g) \rightarrow NO(g) + \tfrac{1}{2} O_2(g)$$

Time/s	[NO$_2$]/mol dm^{-3}
0.0	0.100
5.0	0.017
10.0	0.0090
15.0	0.0062
20.0	0.0047

(a) What is the order of the reaction with respect to nitrogen dioxide?

(b) Calculate the value of the rate constant and give its units.

Solution

(a) Plotting the concentration against time curve gives us the graph shown in Figure 5.8a. This shows a relatively steep decrease in concentration at the start of the reaction with a slower decrease as time goes on. This is characteristic of second-order decay. However, it doesn't prove that the reaction is second order.

In order to prove that the reaction is second order in NO$_2$ we need to use the integrated rate equation for a second-order reaction:

$$\frac{1}{[NO_2]_t} = \frac{1}{[NO_2]_0} + kt$$

If a plot of $\frac{1}{[NO_2]}$ against time, t, is a straight line then we have proved that this is a second-order reaction. The values of $\frac{1}{[NO_2]}$ must be calculated for every data point and so the data table becomes:

Time/s	[NO$_2$]/mol dm^{-3}	1/[NO$_2$]/mol^{-1} dm^3
0.0	0.1	10.0
5.0	0.017	58.8
10.0	0.009	111
15.0	0.0062	161
20.0	0.0047	213

(a)

(b)

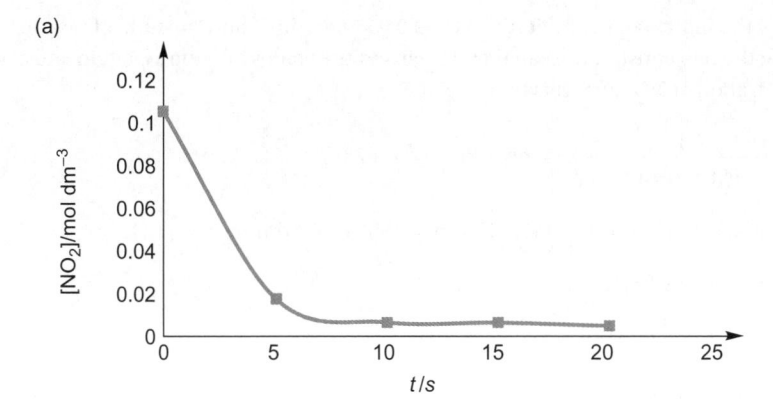

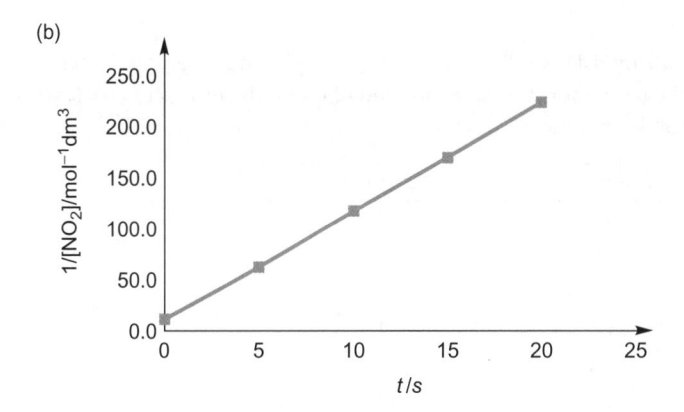

Figure 5.8 (a) [NO$_2$] plotted against time. (b) 1/[NO$_2$] plotted against time.

The plot shown in Figure 5.8b gives a straight line proving that the reaction is second order in NO$_2$.

(b) In this case the gradient of the plot is positive and because the line has the equation of a straight line, the gradient is equal to k, the rate constant.

We can obtain the gradient by taking the difference in y values divided by the difference in x values. This can be done manually if the graph is drawn by hand or it can be obtained from the equation for the line if the graph is plotted using Excel. Either way the gradient can be shown to be 10.2 mol^{-1} dm^3 s^{-1}. As the gradient is equivalent to the rate constant k in a second-order reaction this is also the value of the rate constant.

Worked example 5.4D

If a flask that initially contains 0.056 mol dm^{-3} NO$_2$ is heated at 300 °C, what will be the concentration of NO$_2$ after 1.0 h if the reaction follows second-order kinetics? How long will it take for the concentration of NO$_2$ to decrease to 10% of the initial concentration? The rate constant for the reaction at this temperature is 0.54 mol^{-1} dm^3 s^{-1}.

Solution

The second-order integrated rate equation is: $\dfrac{1}{[NO_2]_t} = \dfrac{1}{[NO_2]_0} + kt$

We know the initial concentration of NO$_2$ is 0.056 mol dm^{-3} and the time of reaction is one hour. Using the rate constant we can therefore insert the values into the equation and calculate the concentration of NO$_2$ after this time.

$$\frac{1}{[NO_2]_t} = \frac{1}{0.056 \text{ mol dm}^{-3}} + 0.54 \text{ mol}^{-1} \text{ dm}^3 \text{ s}^{-1} \times 3600 \text{ s}$$

$$= 17.86 \text{ mol}^{-1} \text{ dm}^3 + 1944 \text{ mol}^{-1} \text{ dm}^3 = 1962 \text{ mol}^{-1} \text{ dm}^3$$

Taking the reciprocal gives:

$[NO_2]_t = 5.10 \times 10^{-4}$ mol dm^{-3}

→ Don't forget to convert the time in hours to seconds by multiplying by 60 × 60.

Worked example 5.4E

An experiment was carried out to investigate the reaction of hex-1-ene with I$_2$.

$$C_6H_{12} + I_2 \rightarrow C_6H_{12}I_2$$

The dependence of rate on [I$_2$] is studied using a large excess of hex-1-ene. Results are recorded below. The data in the table show the concentration of I$_2$ and the product C$_6$H$_{12}$I$_2$ at a series of times after the start of the reaction.

t/s	[I$_2$]/mM	[C$_6$H$_{12}$I$_2$]/mM
0	40.0	0
1000	31.2	8.8
2000	25.6	14.4
3000	21.8	18.2
4000	18.8	21.2
5000	16.6	23.4
6000	15.0	25.0
7000	13.6	26.4
8000	12.4	27.6

→ When one of the reactants is in large excess such as in this example, the concentration of that reactant can be ignored and the rate equation said to be pseudo-first order or second order in the other reactant. In this case we can assume that the kinetics are pseudo-second order in [I$_2$].

(a) By plotting a suitable graph, confirm that the reaction is second order with respect to I$_2$, write the pseudo-second-order rate equation, and calculate the pseudo-second-order rate constant.

(b) At $t = 20\,000$ s, calculate [I$_2$], and [C$_6$H$_{12}$I$_2$].

Solution

(a) Because the reaction is carried out with a large excess of hexene the hexene concentration can be assumed to be constant throughout the reaction. Thus the rate equation can be written as: rate of reaction $= -\frac{d[I_2]}{dt} = k'[I_2]^n$ where k' is the pseudo-second-order rate constant. The question implies that the rate is second order with respect to iodine and so we can use the second-order integrated rate equation: $\frac{1}{[I_2]_t} = \frac{1}{[I_2]_0} + k't$.

We need to construct a column of data to calculate the $\frac{1}{[I_2]}$ values:

t/s	[I$_2$]/mM	1/[I$_2$]/M^{-1}
0	40.0	25.0
1000	31.2	32.1
2000	25.6	39.1
3000	21.8	45.9
4000	18.8	53.2
5000	16.6	60.2
6000	15.0	66.7
7000	13.6	73.5
8000	12.4	80.6

So the graph to plot is $\frac{1}{[I_2]}$ on the y axis against t on the x axis, as shown in Figure 5.9.

→ Ensure you convert the iodine concentration from mM to M by multiplying by 10^{-3}.

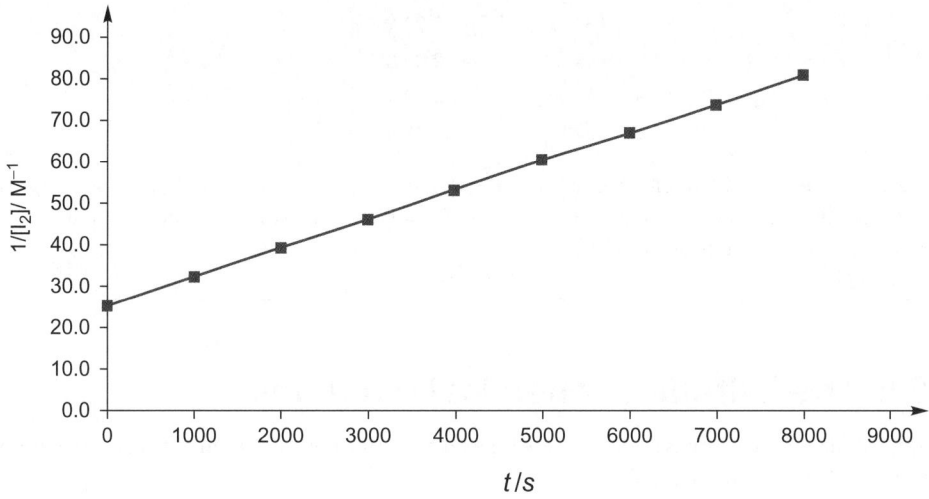

Figure 5.9 Plot of $\frac{1}{[I_2]}$ against time.

The pseudo-second-order rate constant, k', is equal to the gradient of the line. This can be obtained by using the formula: gradient $= \frac{y_2 - y_1}{x_2 - x_1}$.

Substituting values in this equation we find: $k' = \frac{(80.6 - 25.0) \text{ M}^{-1}}{(8000 - 0) \text{ s}} = 6.95 \times 10^{-3} \text{ M}^{-1} \text{ s}^{-1}$.

→ It is more accurate to determine the gradient from your graph by measuring the change in y and dividing by the change in x.

Alternatively this can be obtained from the Excel plot where the gradient of the line is calculated to be 6.90×10^{-3} M^{-1} s^{-1}.

(b) Once we have a value for the pseudo-second-order rate constant, k', we can use the integrated rate equation to find the iodine concentration after a reaction time of 20 000 s.

$$\frac{1}{[I_2]_t} = \frac{1}{40 \times 10^{-3} \text{ M}} + 6.95 \times 10^{-3} \text{ M}^{-1} \text{ s}^{-1} \times 20\,000 \text{ s}$$

$$\frac{1}{[I_2]} = 25 \text{ M}^{-1} + 139 \text{ M}^{-1} = 164 \text{ M}^{-1}$$

→ The inverse of a number is 1 divided by the number, so 1/164 M^{-1} in this case. The units are also inverted and so become M here.

So, taking the inverse of this number:

$[I_2] = 6.1 \times 10^{-3}$ M or 6.1 mM

To answer the second part of the question and work out the concentration of $C_6H_{12}I_2$ formed we need to refer back to the stoichiometric equation. This tells us that one mole of iodine gives one mole of $C_6H_{12}I_2$.

After 20 000 s we have 6.1 mM I_2 left so the amount that has reacted is given by:

$[I_2]_0 - [I_2]_t$ = 40.0 mM – 6.1 mM = 33.9 mM.

From the stoichiometric equation we can see that 33.9 mM of $C_6H_{12}I_2$ must therefore have been formed.

> ### Question 5.13
>
> The following data were obtained for the dimerization of C_4H_6 to C_8H_{12} in a 1 dm³ vessel that proceeds according to the equation:
>
> $2\ C_4H_6\ (g) \rightarrow C_8H_{12}\ (g)$
>
> | time/s | C_4H_6/M |
> |---|---|
> | 0 | 0.0100 |
> | 1000 | 0.00625 |
> | 1800 | 0.00476 |
> | 2800 | 0.00370 |
> | 3600 | 0.00313 |
> | 4400 | 0.00270 |
> | 5200 | 0.00241 |
> | 6200 | 0.00208 |
>
> Determine the order of the reaction with respect to C_4H_6 by graphical methods. Use an appropriate plot to determine the value of the rate constant. Determine the amount in moles of C_8H_{12} formed at time $t = 5000$ s.

5.5 The half-life of chemical reactions

The half-life of a reaction is defined as the time taken for the concentration of a reactant to drop by a half. It is given the symbol $t_{1/2}$.

Half-life in a zero-order reaction

Figure 5.10 shows a plot of concentration against time for a zero-order reaction with the first two half-lives marked. The half-life can be seen to be decreasing as the reaction proceeds.

We can see how the half-life varies during the reaction by taking the integrated rate equation of a zero-order reaction and deriving an expression for the half-life when the initial concentration of reactant, $[A]_0$, drops by half to $[A]_0/2$.

For a zero-order equation: $[A]_t = [A]_0 - kt$

If $[A]_t = [A]_0/2$ the equation becomes: $[A]_0/2 = [A]_0 - kt_{1/2}$

Rearranging to obtain an expression for $t_{1/2}$ gives: $kt_{1/2} = [A]_0 - [A]_0/2 = [A]_0/2$.

Therefore $t_{1/2} = [A]_0/2k$.

So the half-life depends upon the initial concentration of reactant; as this decreases through the reaction, so the half-life decreases.

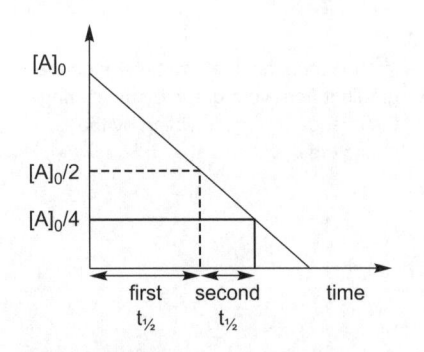

Figure 5.10 Consecutive half-lives in a zero-order reaction.

Half-life in a first-order reaction

Because the concentration of a reactant in a first-order reaction decreases exponentially (as in radioisotope decay) the half-life of a first-order reaction is a constant throughout the reaction as shown in Fig 5.11.

5.5 THE HALF-LIFE OF CHEMICAL REACTIONS

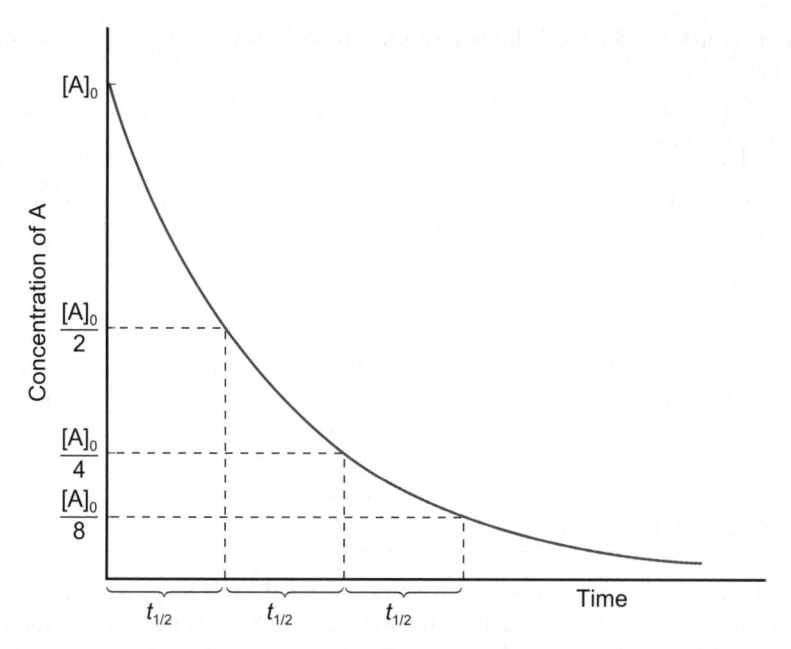

Figure 5.11 The half-life in a first-order reaction is a constant.
Reproduced from Burrows et al, *Chemistry*³ second edition (Oxford University Press, 2013). © Andrew Burrows, John Holman, Andrew Parsons, Gwen Pilling, and Gareth Price 2013.

Using the integrated rate equation we can derive an expression of the half-life of a first-order reaction and show that it depends only upon the rate constant of the reaction.

$$\ln [A]_t = \ln [A]_0 - kt$$

After one half-life, $t_{1/2}$, the concentration of A can be represented by $\frac{[A]_0}{2}$ and so the equation becomes:

$$\ln \frac{[A]_0}{2} = \ln [A]_0 - kt_{1/2}$$

Rearranging to collect the concentration terms together:

$$\ln \frac{[A]_0}{2} - \ln [A]_0 = - kt_{1/2}$$

$$\ln \frac{[A]_0}{2[A]_0} = - kt_{1/2}$$

so, $\ln \frac{1}{2} = - kt_{1/2}$

or $\ln 2 = kt_{1/2}$

so, $t_{1/2} = \frac{\ln 2}{k}$

➔ $\ln[A] - \ln[B] = \ln \frac{[A]}{[B]}$.

➔ If $\ln \frac{1}{A} = -x$ then $\ln A = +x$.

Half-life in a second-order reaction

The integrated rate equation for a second-order reaction is given by:

$$\frac{1}{[A]_t} = \frac{1}{[A]_0} + kt$$

To find an expression for the half-life in a second-order reaction we can again substitute $[A]_t = [A]_0/2$:

$$\frac{1}{[A]_0/2} = \frac{1}{[A]_0} + kt_{1/2}$$

$$kt_{1/2} = \frac{1}{[A]_0/2} - \frac{1}{[A]_0} = \frac{2}{[A]_0} - \frac{1}{[A]_0} = \frac{1}{[A]_0}$$

so, $t_{1/2} = \frac{1}{k[A]_0}$

So the half-life in a second-order reaction depends upon the starting concentration of a reactant. As the reactant concentration drops through the reaction so the half-life will increase and is therefore not constant as in a first-order reaction.

Worked example 5.5A

3.0 g of substance A decomposes and the mass of unreacted A remaining after a certain length of time is found to be 0.375 g. If the half-life of this reaction is 36 minutes, calculate the length of time elapsed if the reaction follows **first-order** kinetics.

Solution

In this question we don't have a value for the rate constant so we can't simply use the equation which relates half-life and rate constant.

There are actually two ways of addressing this type of question. The first uses trial and error and can be useful when the numbers involved are relatively straightforward. We will try this method first.

We know that the starting quantity is 3.0 g. Therefore after one half-life 1.5 g remains. If the concentration decreases by one half at each half-life then the decay can be shown as:

3.0 g → 1.5 g → 0.75 g → 0.375 g

so we can see that after a total of three half-lives the amount of unreacted A is 0.375 g. As one half-life is equivalent to 36 minutes, the total time the process takes is 108 minutes.

The second method uses the integrated rate equation and involves inserting the values for A at the start and the end of the reaction and calculating the time. However, to do this we need the rate constant, k.

$$\ln[A]_t = \ln[A]_0 - kt$$

We can get an expression for k from $t_{1/2} = \frac{\ln 2}{k}$. If we rearrange then $k = \frac{\ln 2}{t_{1/2}}$.

So $k = \ln 2/36$ min $= 0.0193$ min^{-1} $= 1.93 \times 10^{-2}$ min^{-1}.

Now we can use the integrated rate equation and insert the concentration values:

$\ln 0.375 = \ln 3 - 1.93 \times 10^{-2}$ min$^{-1} \times t$

$\ln 0.375 - \ln 3 = -1.93 \times 10^{-2}$ min$^{-1} \times t$

$(\ln 0.375 - \ln 3)/1.93 \times 10^{-2}$ min$^{-1} = -t$

$-2.079/1.93 \times 10^{-2}$ min$^{-1} = -t$

$t = 1.08 \times 10^2$ min $= 108$ min

Worked example 5.5B

Consider the zero-order reaction P → Q in which P molecules are converted to Q molecules. At a certain time, $t = 4$ minutes, there are 16 mg P and after a further six minutes there are 4 mg P remaining. Calculate the time at which there are 8 mg P in the reaction mixture.

5.5 THE HALF-LIFE OF CHEMICAL REACTIONS

Solution

The integrated rate equation for a zero-order reaction is given by: $[A]_t = [A]_0 - kt$. We can use the data in the question to obtain a value for k, the rate constant, and then use this rate constant to find the time at which there are 8 mg P remaining:

4 mg = 16 mg − $k \times$ 6 min

6k min = 16 mg − 4 mg = 12 mg

Therefore $k = (12/6)$ mg min^{-1} = 2 mg min^{-1}.

Substituting the value for k in the integrated rate equation, with the $[A]_t = 8$ mg gives:

8 mg = 16 mg − 2 mg min$^{-1} \times t$

$$t = \left(\frac{16 \text{ mg} - 8 \text{ mg}}{2 \text{ mg min}^{-1}}\right) = 4 \text{ min}$$

So the time at which there are 8 mg P remaining is 4 + 4 = 8 minutes.

➔ Note that the question tells us the concentration of P at $t = 4$ minutes, not $t = 0$ and so we have to add four minutes to the time we have calculated.

Worked example 5.5C

The decomposition of dinitrogen pentoxide at 65 °C:

2 N$_2$O$_5$ (g) → 4 NO$_2$ (g) + O$_2$ (g)

follows the rate equation:

Rate of reaction = k [N$_2$O$_5$ (g)]

In an experiment to investigate the reaction the concentration of N$_2$O$_5$ was measured with time and the following results obtained.

Time/s	[N$_2$O$_5$]/mol dm^{-3}
0	10 × 10^{-2}
0.5	8.6 × 10^{-2}
1.0	7.3 × 10^{-2}
1.5	6.3 × 10^{-2}
2.0	5.4 × 10^{-2}
2.5	4.6 × 10^{-2}
3.0	3.9 × 10^{-2}
3.5	3.4 × 10^{-2}
4.0	2.9 × 10^{-2}

(a) Plot [N$_2$O$_5$] against time.

(b) Plot a second graph to show that the reaction is first-order in N$_2$O$_5$.

(c) Determine the rate constant for the reaction.

(d) Find a value for the half-life for the reation.

Solution

(a) The graph, shown in Figure 5.12, is a simple plot of concentration of N$_2$O$_5$ against time.

(b) To show the reaction is first-order we should use the integrated rate equation. This involves plotting ln[N$_2$O$_5$] against time, so we must first calculate the natural log values and add to the table.

→ Always make sure you include a suitable title on your graphs and label both axes, remembering to add the units.

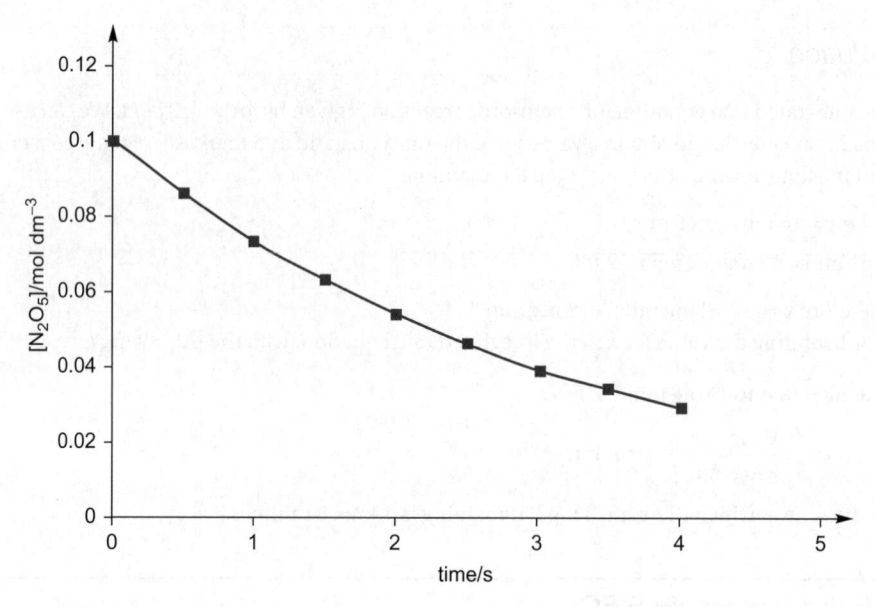

Figure 5.12 Plot of [N_2O_5] against time.

→ Note that as [N_2O_5] gets smaller the value of the ln term gets more negative.

Time/s	[N_2O_5]/mol dm^{-3}	ln[N_2O_5]
0	0.1	−2.30
0.5	0.086	−2.45
1	0.073	−2.62
1.5	0.063	−2.76
2	0.054	−2.92
2.5	0.046	−3.08
3	0.039	−3.24
3.5	0.034	−3.38
4	0.029	−3.54

→ Note that the values of ln[N_2O_5] on the y axis are all negative so this axis is below the line y = 0.

The plot of ln[N_2O_5] is shown in Figure 5.13.

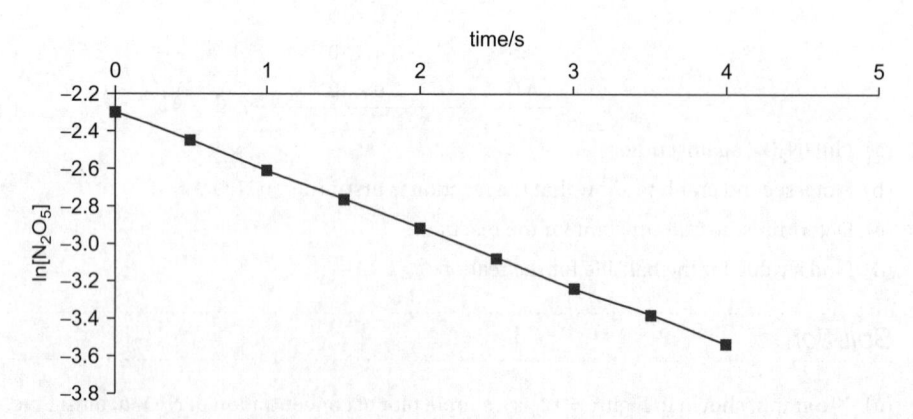

Figure 5.13 Plot of ln[N_2O_5] against time.

The fact that a plot of ln[N_2O_5] against time gives a straight line indicates that the reaction is first order in N_2O_5.

(c) The equation of the line is given by: $\ln[N_2O_5]_t = \ln[N_2O_5]_0 - kt$. The gradient of the line is therefore equal to $-k$, the rate constant.

The gradient is obtained by taking $\frac{y_2 - y_1}{x_2 - x_1}$. This can be done by using your graph and measuring the y change and equivalent x change, remembering to include units as appropriate or by taking values such as: $\frac{-3.54 - (-2.3)}{(4-0)\,s} = -0.31\,s^{-1}$ so $k = 0.31\,s^{-1}$. Alternatively you can obtain the gradient from the equation of the line if you have plotted the graph using Excel.

(d) Once we have the rate constant we can find a value for the half-life, $t_{1/2} = \frac{\ln 2}{0.31\,s^{-1}} = 2.2\,s$.

This can be confirmed by your first plot of $[N_2O_5]$ against time. The initial concentration is 1.0×10^{-2} mol dm^{-3}. The time taken for the concentration to drop by a half to 0.5 mol dm^{-3} is approximately 2.2 s as seen on the graph.

→ Note that the slope of the line is negative. From the integrated rate equation the gradient is equivalent to $-k$, so this will give the expected positive value for the rate constant, k. Like rates, rate constants must be positive.

→ It is always more accurate to measure the y and x change from the graph and obtain the gradient in this way rather than by taking two data points.

Determining the order of a chemical reaction and the rate constant from experimental measurements

If you are given a set of rate data, or if you have taken measurements and have a set of data for a reaction relating concentration of a reactant with time, it is not always obvious what the order of reaction is with respect to the reactant concerned. The flow chart shown in Figure 5.14 suggests a set of steps you can take to determine the order of reaction graphically. It is important to have a procedure for this to avoid having to plot several graphs which is time consuming before deciding the correct one to give you the rate constant.

Question 5.14

A 64 mg initial concentration of a radioactive sample decays by first-order reaction kinetics. At $t = 10$ minutes, the mass of the sample remaining is 16 mg. How many milligrams of this sample remain after 15 minutes?

Question 5.15

The decomposition of sulfuryl chloride, SO_2Cl_2, is a first-order process with a rate constant at 660 K of $4.5 \times 10^{-2}\,s^{-1}$.

(a) If the initial SO_2Cl_2 pressure is 450 torr, what is the partial pressure of the substance after 80 s?

(b) At what time will the partial pressure of SO_2Cl_2 decline to one quarter of its initial value?

Question 5.16

The following data were obtained for the decomposition of $CH_3N_2CH_3$ at 573 K.

Time/s	$[CH_3N_2CH_3] \times 10^3$/mol dm^{-3}
0	1.60
240	1.12
360	0.94
540	0.72
780	0.50
1020	0.35

(a) By inspection of the data determine whether the reaction has first-order kinetics.
(b) Show graphically that the reaction is first order in $CH_3N_2CH_3$.
(c) Determine the rate constant for the reaction.

→ When data are presented in a table in this way the column title indicates that the variable values—in this case $[CH_3N_2CH_3]$—have been multiplied by 10^3. So $[CH_3N_2CH_3]$ at $t = 0$ is 1.6×10^{-3} mol dm^{-3}.

5 REACTION KINETICS

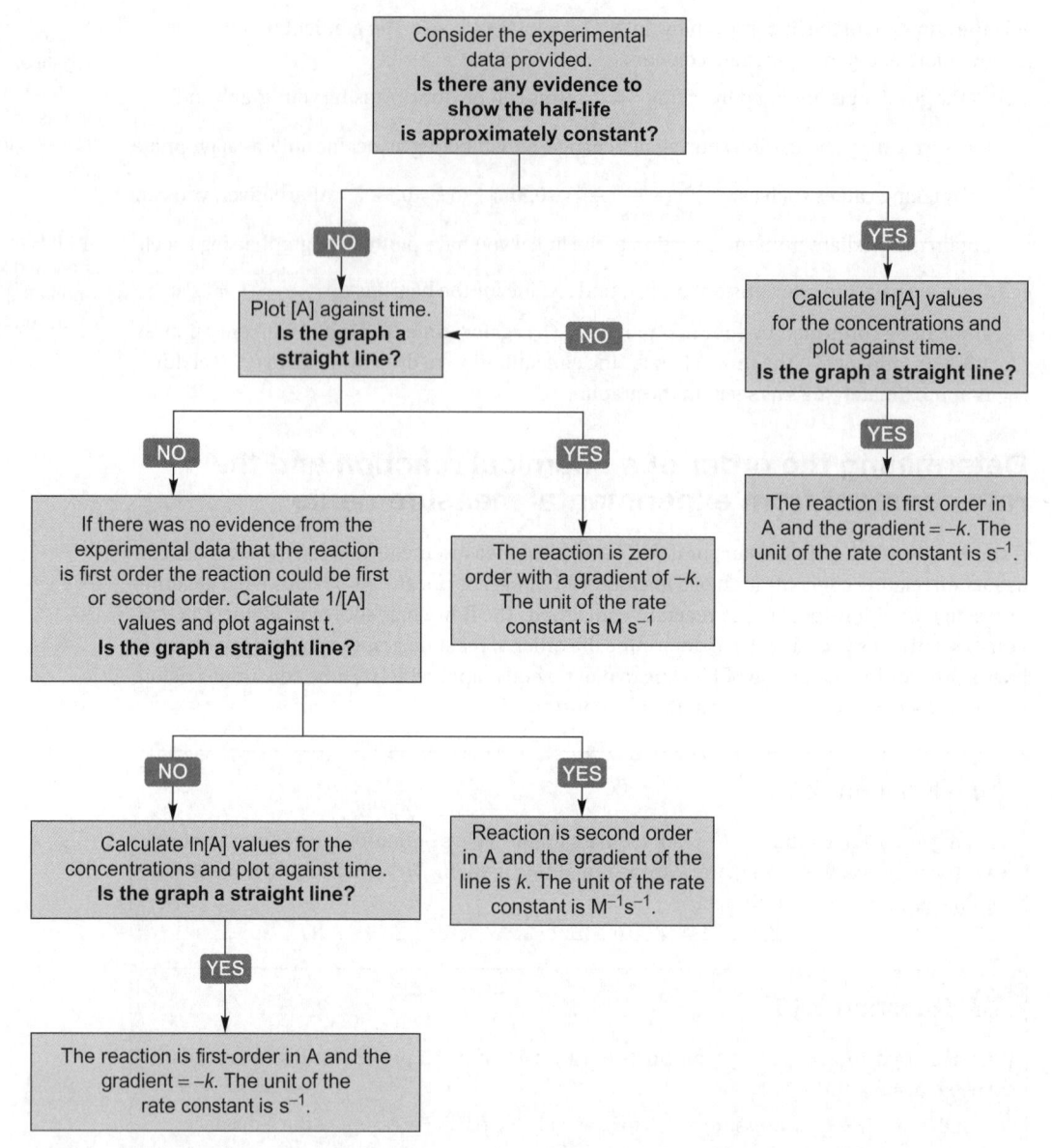

Figure 5.14 Flow chart for determining the order of a chemical reaction graphically from measurements of rates of reaction.

5.6 Dependence of rate of reaction upon temperature — the Arrhenius equation

Chemical reaction can only take place when certain conditions are fulfilled:

- The reactants must collide together (in a reaction which has a molecularity greater than 1).
- They must collide in the correct orientation.
- They must have enough kinetic energy to overcome the activation energy for the reaction.

The distribution of kinetic energies of gaseous molecules in a sample follows a Boltzmann distribution. For any given reaction the activation energy, E_a, is a fixed value and is independent of temperature. In Figure 5.15 this energy distribution would be represented by the line at temperature T_1.

By increasing the temperature to T_2 more of the molecules will move with higher speeds and will therefore possess the activation energy. This means that there are more molecules that are capable of successful reaction. This in turn leads to an increase in the rate of reaction. A rough

5.6 DEPENDENCE OF RATE OF REACTION UPON TEMPERATURE—THE ARRHENIUS EQUATION

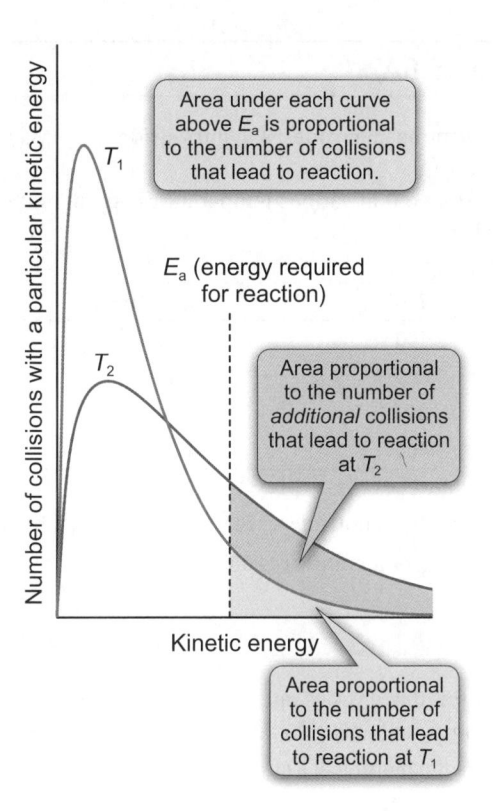

Figure 5.15 Boltzmann distribution of energies in a gas at two temperatures T_1 and T_2 ($T_2 > T_1$). Reproduced from Burrows et al, *Chemistry*[3] second edition (Oxford University Press, 2013). © Andrew Burrows, John Holman, Andrew Parsons, Gwen Pilling, and Gareth Price 2013.

rule of thumb is that increasing the temperature by 10 K results in an approximate doubling of the reaction rate.

The Arrhenius equation relates the rate constant for the reaction to the activation energy and temperature. The rate constant is proportional to the rate of reaction, so an increase in rate constant leads to an increase in rate of reaction.

This can be expressed as: $k = A\,e^{-E_a/RT}$.

If we take natural logs (ln) of both sides of this equation we obtain the following form of the Arrhenius equation: $\ln k = \ln A - \dfrac{E_a}{RT}$

The variables in the equation are:

k = the rate constant (units depend upon order of the reaction).

A = Arrhenius factor or frequency factor or pre-exponential factor—this is a property of the alignment of the reacting molecules.

R = gas constant = 8.314 J K^{-1} mol^{-1}.

T = temperature (K).

This form of the equation is particularly useful as it has the equation of a straight line:

$\ln k = \ln A - \dfrac{E_a}{RT}$ which is analogous to $y = c + mx$.

So by plotting $\ln k$ against $1/T$ we would expect to get a straight line with gradient $-E_a/R$ and intercept on the y axis (i.e. where $1/T = 0$) of $\ln A$.

The importance of this equation is that by measuring the rate of a reaction at different temperatures we can calculate values for the rate constant at these temperatures and can use the plot to obtain a value for the activation energy, E_a. A knowledge of the activation energy for a reaction is extremely useful as it helps us to predict how likely it is for a reaction to occur given a certain set of conditions.

Worked example 5.6A

In the decomposition of nitrogen dioxide, NO_2, nitric oxide and oxygen are formed as products:
$2 NO_2 (g) \rightarrow 2 NO (g) + O_2 (g)$

The reaction was carried out at various temperatures and the rate constants for each temperature derived as follows:

T/°C	k/mol^{-1} dm^3 s^{-1}
100	1.10×10^{-9}
200	1.80×10^{-8}
300	1.20×10^{-7}
400	4.40×10^{-7}

(a) Use a graph to find the activation energy for the reaction and the Arrhenius factor.
(b) Predict a value for k at a temperature of 500 °C.

Solution

(a) A question such as this that gives us rate constant data at different temperatures can be solved by plotting a graph using the Arrhenius equation. The form of the Arrhenius equation that allows the plotting of a straight line is: $\ln k = \ln A - \frac{E_a}{RT}$.

$\ln k$ should be plotted on the y axis and $1/T$ on the x axis. The first step is to calculate values of $\ln k$ and $1/T$ so the data table should be extended:

> Because the Arrhenius equation uses absolute values of temperature we first have to convert °C to K by adding 273.

> Note that ln k has no units and that the values of natural logarithms are all negative when the value of k is less than one.

T/°C	T + 273/K	1/T/K^{-1}	k/mol^{-1} dm^3 s^{-1}	ln k
100	373	0.00268	1.10×10^{-9}	−20.63
200	473	0.00211	1.80×10^{-8}	−17.83
300	573	0.00175	1.20×10^{-7}	−15.94
400	673	0.00149	4.40×10^{-7}	−14.64

A graph can now be plotted of $\ln k$ against $1/T$, as shown in Figure 5.16.

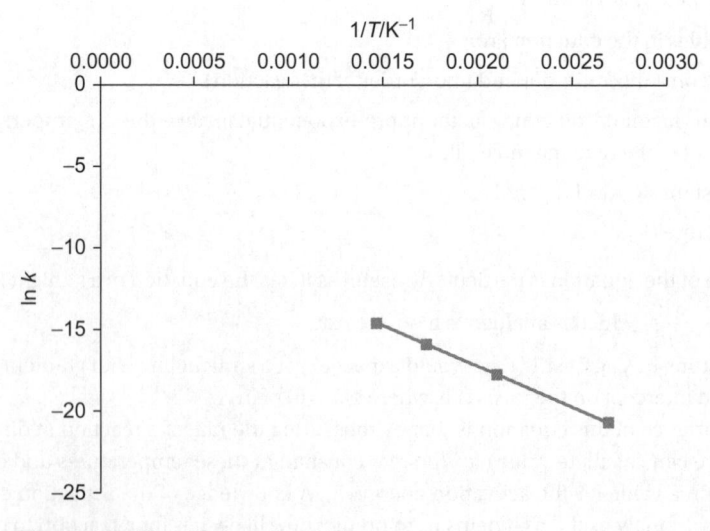

Figure 5.16 Plot of ln k against 1/T.

5.6 DEPENDENCE OF RATE OF REACTION UPON TEMPERATURE – THE ARRHENIUS EQUATION

The plot is a straight line with a negative gradient. When $1/T$ is plotted on the x axis the gradient is equal to $-E_a/R$.

The next step is to obtain the gradient, either by measuring distances along x and y on your graph or using the expression: gradient $=\dfrac{y_2-y_1}{x_2-x_1}$.

Substituting values into this expression gives:

$$\dfrac{-14.64-(-20.63)}{(0.00149-0.00268)\,\text{K}^{-1}}=\dfrac{5.99}{-0.00119\,\text{K}^{-1}}=-5.03\times 10^3\,\text{K}.$$

From the Arrhenius equation the gradient is equal to $-\dfrac{E_a}{R}$

So: $-\dfrac{E_a}{R}=-5.03\times 10^3\,\text{K}$

R, the gas constant, has a value of $8.314\,\text{J K}^{-1}\,\text{mol}^{-1}$:

So $-E_a = -5.03\times 10^3\,\text{K}\times 8.314\,\text{J K}^{-1}\,\text{mol}^{-1} = -41.8\,\text{kJ mol}^{-1}$

So $E_a = 41.8\,\text{kJ mol}^{-1}$

> Note that there are no units for the numerator (top line value) as it is a logarithm.

The Arrhenius factor is represented in the Arrhenius equation as $\ln A$ and is equivalent to the point at which the line crosses the x axis. So to find the Arrhenius factor the line must be extended back to the y axis where $x=0$. In order to do this we need to plot the graph with both axes shown drawn to the origin. This plot is shown in Figure 5.17. By extending the line to the y axis we find that it crosses at about $y=-7.2$. This can be done if you are using a spreadsheet to plot your graph by setting x equal to zero in the equation for the line as can be seen in Figure 5.17.

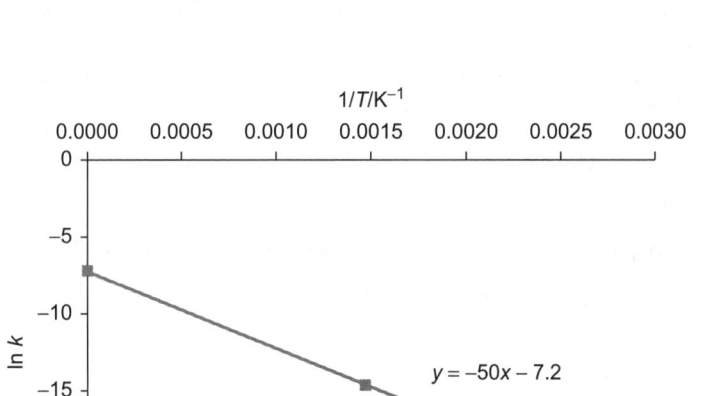

Figure 5.17 Extrapolating Figure 5.16 to $1/T = 0$ to find the value of A.

> Extrapolating back to $x = 0$ is very hard to do on a hand-drawn graph.

If $\ln A = -7.2$, $A = 7.5\times 10^{-4}\,\text{mol}^{-1}\,\text{dm}^3\,\text{s}^{-1}$

Although $\ln A$ has no units, A itself has the same units as the rate constant.

(b) Once we have a value for the activation energy, E_a, and the Arrhenius factor we can use these in the Arrhenius equation to obtain a value of k at the higher temperature of 500 °C.

The equation is: $\ln k = \ln A - \dfrac{E_a}{RT}$

So inserting values of $T = 500 + 273\,\text{K} = 773\,\text{K}$, $E_a = 41.50\,\text{kJ mol}^{-1}$ and $A = 7.62\times 10^{-4}\,\text{mol}^{-1}\,\text{dm}^3\,\text{s}^{-1}$

$\ln k = \ln 7.62\times 10^{-4}\,\text{mol}^{-1}\,\text{dm}^3\,\text{s}^{-1} - 41.50\times 10^3\,\text{J mol}^{-1}/8.314\,\text{J K}^{-1}\,\text{mol}^{-1}\times 773\,\text{K}$

$= -7.18 - 6.47 = -13.65$

So $k = 1.18\times 10^{-6}\,\text{mol}^{-1}\,\text{dm}^3\,\text{s}^{-1}$

As the value of A is difficult to obtain accurately it is preferable to carry out a calculation using two data points and cancelling the ln A value as shown. By subtracting the equations at the two data points and collecting like terms we obtain:

$$\ln\left(\frac{k_1}{k_2}\right) = \frac{E_A}{R}\left(\frac{1}{T_2} - \frac{1}{T_1}\right)$$

Use the temperature of 400 °C as the exact data point (T_2) calculate the value of k_1 at the new temperature of 500 °C (T_1).

$$\ln\left(\frac{k_1}{4.40 \times 10^{-7}}\right) = \frac{4.18 \times 10^4 \text{ J mol}^{-1}}{8.314 \text{ J k}^{-1}\text{ mol}^{-1}}\left(\frac{1}{673 \text{ K}} - \frac{1}{773 \text{ K}}\right) = 0.966$$

So $k_1 = 1.16 \times 10^{-6} \text{ mol}^{-1} \text{ dm}^3 \text{ s}^{-1}$

Worked example 5.6B

The reaction $C_2H_6(g) \to 2\ CH_3\cdot(g)$ was monitored and it was found that at 400 K the rate constant $k = 0.052 \text{ s}^{-1}$ and at 550 K the rate constant was 0.54 s^{-1}. Calculate the activation energy for the reaction.

Solution

In this problem we have two data points for the rate constant and the temperature. This isn't enough data to plot a graph but we can use the Arrhenius equation to set up two simultaneous equations to solve for the activation energy.

Taking the Arrhenius equation, $\ln k = \ln A - \frac{E_a}{RT}$, we can write two separate equations with the data provided:

→ Here we have two equations with two unknowns so we need to cancel one of the unknowns and then solve for the other.

$$\ln 0.052 = \ln A - E_a/(R \times 400 \text{ K}) \quad (5.2)$$

$$\ln 0.54 = \ln A - E_a/(R \times 550 \text{ K}) \quad (5.3)$$

If we subtract equation 5.3 from equation 5.2 the ln A values will cancel:

$$\ln 0.052 - \ln 0.54 = -E_a/(R \times 400 \text{ K}) - (-E_a/(R \times 550 \text{ K}))$$

This becomes:

→ To obtain the left hand side of the equation we have used the relationship that $\ln A - \ln B = \ln \frac{A}{B}$. To obtain the right hand side of the equation we have collected the like terms $\frac{E_a}{R}$ together outside the bracket leaving the inverse temperature terms to be calculated.

$$\ln\frac{0.052}{0.54} = \frac{E_a}{R}\left(-\frac{1}{400 \text{ K}} + \frac{1}{550 \text{ K}}\right)$$

Working out the numerical terms on both sides gives us:

$$\ln 0.0963 = -\frac{E_a}{R} \times 0.682 \times 10^{-3} \text{ K}^{-1}$$

Rearranging to give an expression for E_a gives:

$$E_a = -\ln 0.0963 \times 8.314 \text{ J K}^{-1} \text{ mol}^{-1}/0.682 \times 10^{-3} \text{ K}^{-1}$$
$$= -(-2.34 \times 8.314 \text{ J K}^{-1} \text{ mol}^{-1}/0.682 \times 10^{-3} \text{ K}^{-1})$$
$$= 28.5 \times 10^3 \text{ J mol}^{-1} = 28.5 \text{ kJ mol}^{-1}$$

> **Question 5.17**
>
> The following set of data was obtained for the conversion of cyclopropane to propene at various temperatures. By plotting an appropriate graph determine the activation energy for the reaction.
>
> | T/°C | k/s^{-1} |
> |------|------------|
> | 480 | 1.8×10^{-4} |
> | 530 | 2.7×10^{-3} |
> | 580 | 3.0×10^{-2} |
> | 630 | 0.26 |
>
> Determine the value of the rate constant at 700 °C.

5.7 Reaction mechanisms and rate dependence

The majority of chemical reactions proceed by a series of several steps called **elementary reactions**. The overall set of elementary reactions that describe the way the reactants are converted into products is called the **reaction mechanism**. The slowest step in the mechanism of a multi-step reaction is known as the **rate-determining step** or **rate-limiting step** as the kinetics of this step determine the overall rate of reaction.

The number of reactant species involved in an elementary reaction is called the **molecularity** of the reaction. An elementary unimolecular reaction obeys a first-order rate equation; an elementary bimolecular reaction obeys a second-order rate equation. In many reaction mechanisms there are highly reactive intermediates whose concentrations are quickly used up as they react rapidly. The **steady-state approximation** assumes that the rate of change of concentration of a reactive species in a reaction mechanism is approximately zero, i.e. the reactive species is used up as fast as it is formed so its concentration stays unchanged with time. If the reactive species is A, the steady-state approximation says: $\frac{d[A]}{dt}=0$. Steady state is reached when the rate of production of an intermediate in one step is the same as its rate of removal in a following step.

Mechanisms with reversible steps

Some reaction mechanisms involve a reversible step. For example the reaction A → C may proceed via an intermediate B:

$$A \underset{k_{-1}}{\overset{k_1}{\rightleftharpoons}} B$$

$$B \xrightarrow{k_2} C$$

The constants k_1, k_{-1}, and k_2 are the rate constants for the respective reactions in the directions shown.

There are two types of kinetics depending upon the relative sizes of the rate constants: pre-equilibrium and short-lived intermediate.

Pre-equilibrium

This type of behaviour occurs when the rate of the second reaction is relatively slow and the first step has time to equilibrate. We can write the equilibrium constant for the first step as: $\frac{[B]}{[A]} = K_c$ where K_c is the equilibrium constant for the first reaction. This can be also expressed as: $\frac{k_1}{k_{-1}} = K_c$.

The concentration of the product B can also be written as:

$$[B] = K_c[A] = \frac{k_1}{k_{-1}}[A]$$

So in this case, because the second step in the reaction is slow it is defined as the rate-determining step and the overall kinetics of the reaction are determined by the rate of formation of B—which is directly linked to the rate of formation of product C. Thus the rate of reaction can be expressed in any of the following ways:

$$\text{rate of reaction} = -\frac{d[A]}{dt} = \frac{d[C]}{dt} = k_2[B] = k_2\frac{k_1}{k_{-1}}[A]$$

Short-lived intermediate

If the second step of the reaction is relatively fast such that B is used up rapidly (in this case either k_{-1} or k_2 is much larger than k_1) then the steady-state approximation can be applied to B. The **steady-state approximation** assumes the concentration of all **intermediates** remains constant and small throughout the reaction. In this case:

$$\frac{d[B]}{dt} = 0 = \text{rate of formation of B} - \text{rate of removal of B}$$

$$\frac{d[B]}{dt} = k_1[A] - (k_{-1}[B] + k_2[B]) = 0$$

So $k_1[A] = (k_{-1}[B] + k_2[B])$

So $[B] = \dfrac{k_1[A]}{(k_{-1} + k_2)}$

So the rate of reaction $= \dfrac{d[C]}{dt} = k_2[B] = \dfrac{k_1 k_2 [A]}{(k_{-1} + k_2)}$

> A short-lived intermediate is any species that does not appear in the overall reaction but which is included in the mechanism.

1. If the conversion of B to C is much faster than the conversion of B back to A then k_2 will be much larger than k_{-1} so the rate becomes:

$$\frac{d[C]}{dt} = k_2[B] = \frac{k_1 k_2 [A]}{k_2} = k_1[A]$$

In this case the rate-determining step is the first step and equilibrium is not established. B is known as a short-lived intermediate.

2. If the conversion of B back to A is much faster than the conversion of B to C then k_{-1} will be much larger than k_2 and the rate of reaction becomes:

$$\frac{d[C]}{dt} = k_2[B] = \frac{k_1 k_2 [A]}{k_{-1}}$$

This is known as the pre-equilibrium case.

Worked example 5.7A

Given the following mechanism for the catalytic destruction of ozone:

$$O_3 \underset{k_{-1}}{\overset{k_1}{\rightleftharpoons}} O_2 + O$$

$$O_3 + O \xrightarrow{k_2} 2\,O_2$$

(a) Determine the equation for the overall reaction.
(b) State the intermediate.

(c) Give the rate equation for each step (in terms of $[O_3]$).

(d) If the empirical rate equation is:

$$-\frac{d[O_3]}{dt} = k\frac{[O_3]^2}{[O_2]}$$

What is the probable rate-determining step (RDS)?

(e) Use the steady-state approximation and assume that the conversion of O back to O_3 occurs much faster than its conversion to O_2 to derive a rate equation consistent with the experimental one.

Solution

(a) Adding together the two steps yields: $2\,O_3 \rightarrow 3\,O_2$.

(b) The intermediate is the O atom since it doesn't appear in the overall reaction.

(c) The rate equations for each step are:

Step 1: forward direction: $-\frac{d[O_3]}{dt} = k_1[O_3]$

Step 1: backward reaction: $\frac{d[O_3]}{dt} = k_{-1}[O_2][O]$

Step 2: $-\frac{d[O_3]}{dt} = k_2[O_3][O]$

(d) The most likely RDS is step 2 since:

i. step 1 is a simple unimolecular reaction;

ii. the reverse of step 1 does not involve ozone.

(e) Assuming the rate-determining step is the second step in the reaction, to derive a rate expression we need to assume the steady-state approximation for the intermediate [O]:

$$\frac{d[O]}{dt} = \text{rate of production of O} - \text{rate of removal of O} = 0$$

$$= k_1[O_3] - (k_2[O_3][O] + k_{-1}[O_2][O]) = 0$$

So: $k_1[O_3] = (k_2[O_3][O] + k_{-1}[O_2][O])$

$k_1[O_3] = [O](k_2[O_3] + k_{-1}[O_2])$

To obtain an expression for the intermediate [O]:

$$[O] = \frac{k_1[O_3]}{(k_2[O_3] + k_{-1}[O_2])}$$

The rate equation for the second step is given by:

$$-\frac{d[O_3]}{dt} = k_2[O_3][O]$$

Substituting for [O] in this equation we obtain:

$$-\frac{d[O_3]}{dt} = k_2[O_3]\frac{k_1[O_3]}{(k_2[O_3] + k_{-1}[O_2])}$$

In the question we are given the information that O is a short lived intermediate and its rate of conversion back to O_3 is much faster than its conversion to O_2 so we can say that $k_{-1} \gg k_2$. We can therefore ignore $k_2[O_3]$ as k_2 is very small and the rate equation becomes:

$$-\frac{d[O_3]}{dt} = k_2[O_3]\frac{k_1[O_3]}{(k_{-1}[O_2])}$$

This can be rearranged to give:

$$-\frac{d[O_3]}{dt} = k_2 \frac{k_1[O_3]^2}{k_{-1}[O_2]} = k\frac{[O_3]^2}{[O_2]}$$

where $k = \dfrac{k_2 k_1}{k_{-1}}$

So the expression agrees with the experimentally determined rate equation of:

$$-\frac{d[O_3]}{dt} = k\frac{[O_3]^2}{[O_2]}$$

Worked example 5.7B

In the acid-catalysed hydrolysis of an ester, E, the reaction scheme can be written as:

$$E + H_3O^+ \underset{k_{-1}}{\overset{k_1}{\rightleftharpoons}} EH^+ + H_2O$$

$$EH^+ + H_2O \xrightarrow{k_2} \text{products}$$

EH^+ is the protonated form of the ester E. Assuming that EH^+ is consumed rapidly and therefore using the steady-state approximation, show that the rate of formation of products is:

Rate = k_{eff} [E][H_3O^+]

Derive an expression for k_{eff} as a combination of the rate constants k_1, k_{-1}, and k_2.

Solution

As step 2 is the slow step we can express the rate of formation of products in this step as:

Rate = $k_2[EH^+][H_2O]$

This requires us to derive an expression for [EH^+] which is an intermediate. Using the steady-state approximation for the rate of change of EH^+ concentration with time we can write:

$$\frac{d[EH^+]}{dt} = \text{rate of production of } EH^+ - \text{rate of removal of } EH^+ = 0$$

From reaction 1 in the forward direction the rate of production of [EH^+] can be expressed by:

$$\frac{d[EH^+]}{dt} = k_1[E][H_3O^+]$$

From reaction 1 in the reverse direction the rate of consumption of [EH^+] can be expressed by:

$$-\frac{d[EH^+]}{dt} = k_{-1}[EH^+][H_2O]$$

In reaction 2 the rate of removal of EH^+ can be written as:

$$-\frac{d[EH^+]}{dt} = k_2[EH^+][H_2O]$$

So applying the steady-state approximation:

$$\frac{d[EH^+]}{dt} = k_1[E][H_3O^+] - (k_{-1}[EH^+][H_2O] + k_2[EH^+][H_2O]) = 0$$

This becomes:

$$k_1[E][H_3O^+] = k_{-1}[EH^+][H_2O] + k_2[EH^+][H_2O] = (k_{-1} + k_2)[EH^+][H_2O]$$

Rearranging to get an expression for [EH^+]:

$$[EH^+] = \frac{k_1[E][H_3O^+]}{(k_{-1} + k_2)[H_2O]}$$

Substituting for [EH$^+$] in the rate equation we obtain:

$$\text{Rate} = k_2 \frac{k_1[\text{E}][\text{H}_3\text{O}^+]}{(k_{-1}+k_2)[\text{H}_2\text{O}]}[\text{H}_2\text{O}]$$

Which becomes:

$$\text{Rate} = \frac{k_1 k_2}{(k_{-1}+k_2)}[\text{E}][\text{H}_3\text{O}^+]$$

So $k_{\text{eff}} = \dfrac{k_1 k_2}{(k_{-1}+k_2)}$

Worked example 5.7C

Suggest a plausible mechanism for the reaction below and derive a rate equation for the proposed mechanism.

$$\text{H}_2(g) + \text{I}_2(g) \rightarrow 2\,\text{HI}(g)$$

Solution

There is no simple answer to this type of question; indeed more than one correct answer may be possible. You are asked to propose a plausible mechanism so, provided you justify your answer, any chemically sensible mechanism would be acceptable. We are given no empirical rate information and so we can't determine whether any one answer is correct.

We have to assume that the reaction proceeds by more than one step. In this case there are at least two possibilities for the first step in the mechanism. Either the H–H or I–I bond could break to give atoms. As the I–I bond is weaker (151 kJ mol^{-1}) than the H–H bond (436 kJ mol^{-1}) we can assume that the I–I bond will break first in a unimolecular elementary step. A possible mechanism would involve the following three-step reaction where k_1, k_2 and k_3 are rate constants for each step:

$\text{I}_2 \rightarrow 2\,\text{I}\;(k_1)$

$\text{I} + \text{H}_2 \rightarrow \text{HI} + \text{H}\;(k_2)$

$\text{H} + \text{I} \rightarrow \text{HI}\;(k_3)$

If we take one of the steps that involves the formation of products we can write a rate equation for the overall reaction:

$$\frac{d[\text{HI}]}{dt} = k_2[\text{I}][\text{H}_2]$$

However, this can't be an empirical rate equation as it includes the term [I] and I atoms are intermediates. To solve for I then we should use the steady-state approximation that I atoms are used up as fast as they are produced. We should therefore write rate equations for the formation and consumption of I and then write the rate equation. The steady-state approximation for [I] states:

$$\frac{d[\text{I}]}{dt} = \text{rate of production of I} - \text{rate of removal of I} = 0$$

I is produced in step 1 according to:

$$\frac{d[\text{I}]}{dt} = k_1[\text{I}_2]$$

and is used up in steps 2 and 3 according to the following:

$$-\frac{d[\text{I}]}{dt} = k_2[\text{I}][\text{H}_2]$$

$$-\frac{d[\text{I}]}{dt} = k_3[\text{H}][\text{I}]$$

So equating the rate of production to the rate of removal we obtain:

$$k_1[I_2] = k_2[I][H_2] + k_3[H][I]$$

From this expression we can get an equation for [I]:

$$[I] = \frac{k_1[I_2]}{k_2[H_2] + k_3[H]}$$

We can now substitute this into the expression for the rate equation we wrote:

$$\frac{d[HI]}{dt} = k_2 \frac{k_1[I_2]}{k_2[H_2] + k_3[H]}[H_2]$$

You will see that this still can't be an acceptable expression for the rate equation as it contains a term in [H]. So we need also to apply the steady-state approximation to [H], another intermediate in our proposed mechanism.

H is produced in step 2 and used up in step 3 so we can write two rate equations:

$$\frac{d[H]}{dt} = k_2[I][H_2]$$

and:

$$-\frac{d[H]}{dt} = k_3[I][H]$$

Equating these two expressions we obtain:

$$k_2[I][H_2] = k_3[I][H]$$

So $[H] = \dfrac{k_2[H_2]}{k_3}$

We can now substitute for [H] in the rate equation:

$$\frac{d[HI]}{dt} = k_2 \frac{k_1[I_2]}{k_2[H_2] + k_3 \dfrac{k_2[H_2]}{k_3}}[H_2]$$

Cancelling the $[H_2]$ and k_3 terms we then obtain:

$$\frac{d[HI]}{dt} = k_2 \frac{k_1[I_2]}{k_2 + k_2} = \frac{k_1[I_2]}{2}$$

So the rate equation can be written as Rate = $k_1 \dfrac{[I_2]}{2}$

An alternative mechanism would be:

$$I_2 \underset{k_{-1}}{\overset{k_1}{\rightleftharpoons}} 2\,I$$

$$H_2 + 2\,I \xrightarrow{k_2} 2\,HI$$

This mechanism assumes that the dissociation of I_2 is reversible—which is more likely—and the I produced reacts with H_2 in a termolecular reaction, which is significantly less likely.

In this mechanism we have only one intermediate, I, and so the steady-state approximation need only be applied to this species as follows:

5.7 REACTION MECHANISMS AND RATE DEPENDENCE

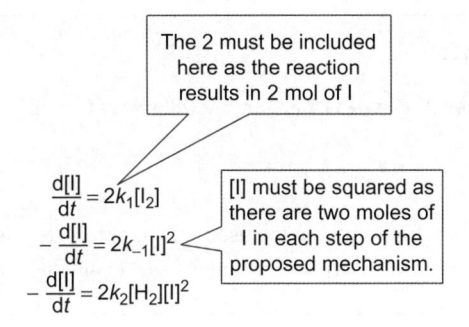

$$\frac{d[I]}{dt} = 2k_1[I_2]$$

The 2 must be included here as the reaction results in 2 mol of I

$$-\frac{d[I]}{dt} = 2k_{-1}[I]^2$$

[I] must be squared as there are two moles of I in each step of the proposed mechanism.

$$-\frac{d[I]}{dt} = 2k_2[H_2][I]^2$$

Therefore equating the rate of production with the rate of consumption of I:

$$2k_1[I_2] = 2k_{-1}[I]^2 + 2k_2[H_2][I]^2$$

And writing an expression for $[I]^2$ we obtain:

$$[I]^2 = \frac{2k_1[I_2]}{2k_{-1} + 2k_2[H_2]}$$

In this second mechanism the only step that leads to products is the second step (as the first is an equilibrium). So we can write an expression for the rate equation based upon this step:

$$\text{Rate} = 2k_2[H_2][I]^2$$

Substituting for $[I]^2$ in this equation we obtain:

$$\text{Rate} = 2k_2[H_2]\frac{2k_1[I_2]}{2k_{-1} + 2k_2[H_2]} = \frac{2k_2k_1[H_2][I_2]}{k_{-1} + k_2[H_2]}$$

If k_{-1} is small (i.e. there is little conversion of I back to I_2) we get the following expression for the rate. This is equivalent to no pre-equilibrium existing:

$$\text{Rate} = \frac{2k_2k_1[H_2][I_2]}{k_{-1} + k_2[H_2]} = 2k_1[I_2]$$

This can be compared with the rate of reaction derived for the previous mechanism:

$$\text{Rate} = k_1\frac{[I_2]}{2}$$ The value of the constant k_1 will be different in each mechanism.

If $k_2[H_2]$ is small compared to k_{-1} (i.e. step 2 is very slow and the equilibrium in step 1 lies to the left) then we obtain:

$$\text{Rate} = \frac{2k_2k_1[H_2][I_2]}{k_{-1} + k_2[H_2]} = \frac{2k_2k_1[H_2][I_2]}{k_{-1}} = k[H_2][I_2] \text{ where } k = \frac{2k_2k_1}{k_{-1}}$$

Since this rate equation involves both reactants it could be argued that this is a one-step mechanism which is first-order in both I_2 and H_2.

On the other hand, if we use a very large excess of H_2 compared to I_2 such that the concentration of H_2 changes only very slightly then $k_2[H_2] \gg k_{-1}$ and the expression becomes:

$$\text{Rate} = \frac{2k_2k_1[H_2][I_2]}{k_{-1} + k_2[H_2]} = 2k_1[I_2].$$ In this case the reaction is said to be **pseudo-first order** with a very large and virtually unchanged concentration of one reactant (H_2) making the rate dependent upon the concentration of the reactant that is present in the lower concentration.

> **Question 5.18**
>
> In the stepwise reaction: A → B → C there are two elementary steps:
>
> $A \xrightarrow{k_1} B$
>
> $B \xrightarrow{k_2} C$
>
> Assuming that B is an intermediate that is used up as rapidly as it is produced apply the steady-state approximation to derive an expression for the rate of formation of C.

> **Question 5.19**
>
> If the reaction:
>
> $2\,N_2O_5 \rightarrow 4\,NO_2 + O_2$
>
> follows the mechanism:
>
> $N_2O_5 \underset{k_b}{\overset{k_f}{\rightleftharpoons}} NO_2 + NO_3$
>
> $NO_3 + NO_2 \rightarrow NO + NO_2 + O_2$ (rate constant k_2)
>
> $NO_3 + NO \rightarrow 2\,NO_2$ (rate constant k_3)
>
> use the steady-state approximation to derive the rate equation.

Turn to the Synoptic questions section on page 174 to attempt questions that encourage you to draw on concepts and problem-solving strategies from several topics within a given chapter to come to a final answer.

Final answers to numerical questions appear at the end of the book, and full worked solutions appear on the book's website, where you can also find a set of bonus questions for each chapter. Go to www.oxfordtextbooks.co.uk/orc/chemworkbooks/.

6
Electrochemistry

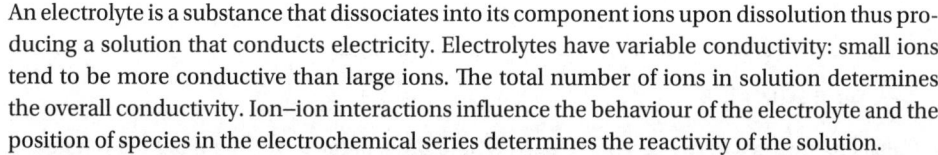

An electrolyte is a substance that dissociates into its component ions upon dissolution thus producing a solution that conducts electricity. Electrolytes have variable conductivity: small ions tend to be more conductive than large ions. The total number of ions in solution determines the overall conductivity. Ion–ion interactions influence the behaviour of the electrolyte and the position of species in the electrochemical series determines the reactivity of the solution.

Electrochemical processes involve an exchange of electrons: oxidation is the loss of electrons and reduction is the gain of electrons. The following section will cover electrostatic interactions, activity, ionic strength, the Debye-Hückel theory, conductivity, electrochemical cells, the Nernst equation, and Faraday's laws.

6.1 Electrostatic interactions

Coulomb's law shows how the force between the ions in solution depends on the nature of the solvent and the distance between the ions:

$$F = \frac{q_{Na^+} q_{Cl^-}}{4\pi\varepsilon_0 \varepsilon_R r^2}$$

where:

- F is the force (N).
- q_{Na^+} is the charge on the sodium cation (C).
- q_{Cl^-} is the charge on the chloride anion (C).
- ε_0 is the permittivity of a vacuum (8.854×10^{-12} C^2 N^{-1} m^{-2}).
- ε_R is the dielectric constant of the medium (a dimensionless quantity).
- r is the separation distance (m).

➔ The dielectric constant varies from one solvent to another and values are available in a data book or textbooks. Water is a good solvent for electrolytes and its dielectric constant is 78.54.

Worked example 6.1A

Calculate the force between a pair of ions, Na$^+$ and Cl$^-$, separated by 2.0 nm in (a) a vacuum and (b) water, and comment on the magnitude of the values and the sign (either positive or negative) of the force.

Solution

Use Coulomb's law to determine the force:

$$F = \frac{q_{Na^+} q_{Cl^-}}{4\pi\varepsilon_0 \varepsilon_R r^2}$$

The charges on the cation and anion are 1.602×10^{-19} C and -1.602×10^{-19} C respectively and the dielectric constant of water is 78.54:

> The charge on the univalent cations and anions is the charge of an electron, the cation has a positive charge whilst the anion has a negative charge. SI units should be used throughout. The dielectric constant of a vacuum is effectively equal to one.

In a vacuum:

$$F = \frac{q_{Na^+} q_{Cl^-}}{4\pi\varepsilon_0 \varepsilon_R r^2} = \frac{(1.602 \times 10^{-19}\ C) \times (-1.602 \times 10^{-19}\ C)}{4 \times 3.142 \times (8.854 \times 10^{-12}\ C^2\ N^{-1}\ m^{-2}) \times (2.0 \times 10^{-9}\ m)^2}$$

$$F = -5.8 \times 10^{-11}\ N$$

In water:

$$F = \frac{q_{Na^+} q_{Cl^-}}{4\pi\varepsilon_0 \varepsilon_R r^2} = \frac{(1.602 \times 10^{-19}\ C) \times (-1.602 \times 10^{-19}\ C)}{4 \times 3.142 \times (8.854 \times 10^{-12}\ C^2\ N^{-1}\ m^{-2}) \times (78.54) \times (2.0 \times 10^{-9}\ m)^2}$$

$$F = -7.3 \times 10^{-13}\ N$$

> The dielectric constant for water is 78.54, the larger the dielectric constant the smaller the force will be between ions.

The force is reduced when the ions are in water compared to being in a vacuum. Ions of opposite charge should attract one another; the negative sign for the force indicates an attractive force.

Question 6.1

Calculate the force between a pair of ions K^+ and Br^- separated by 5.0 nm in (a) a vacuum and (b) ethanol, and comment on the magnitude of the values and the sign (either positive or negative) of the force. The dielectric constant of ethanol is 24.3.

6.2 Activity

The activity of an electrolyte, a_{+-}, is given by:

$$a_{+-} = (a_\pm)^\nu$$

where:

- a_{+-} is the activity of the electrolyte (a dimensionless quantity).
- $a_\pm$ is the mean ionic activity (a dimensionless quantity).
- ν is the total number of ions per electrolyte molecule (a dimensionless quantity).

> Note the similarity between the notation used for the activity of the electrolyte, a_{+-}, and the mean ionic activity, $a_\pm$. Care should be taken to avoid confusion between these terms.

The mean ionic activity, $a_\pm$ is given by:

$$a_\pm = \gamma_\pm m_\pm$$

where:

- $a_\pm$ is the mean ionic activity (a dimensionless quantity).
- $\gamma_\pm$ is the mean ionic activity coefficient (a dimensionless quantity).
- $m_\pm$ is the mean ionic molality (a dimensionless quantity, obtained by dividing the mean molality in mol kg^{-1} by 1 mol kg^{-1}).

Note that the mean ionic molality is the geometric mean of the individual ionic molalities and is determined by:

$$m_\pm = [(m_+)^{\nu^+} (m_-)^{\nu^-}]^{1/\nu}$$

where:

> 'Taking the geometric mean' is a phrase used to describe a particular mathematical action. For example, if two numbers are under consideration then they are multiplied together and the square root is then taken; if three numbers are under consideration then they are multiplied together and the cube root is taken.

- $m_\pm$ is the mean ionic molality (a dimensionless quantity, obtained by dividing the mean molality in mol kg^{-1} by 1 mol kg^{-1}).
- m_+ is the molality of the cation (mol kg^{-1}).
- m_- is the molality of the anion (mol kg^{-1}).
- ν^+ is the number of cations (dimensionless).
- ν^- is the number of anions (dimensionless).
- ν is the total number of ions i.e. $\nu = (\nu^+) + (\nu^-)$ (dimensionless).

Worked example 6.2A

For a 1:1 electrolyte such as NaCl with a molality of m the activity may be determined by considering the number of ions produced upon dissociation, and their molality:

$$NaCl \rightarrow Na^+ + Cl^-$$
$$a_{+-} = (a_\pm)^\nu$$

where:

- a_{+-} is the activity of the electrolyte (a dimensionless quantity).
- $a_\pm$ is the mean ionic activity (a dimensionless quantity).
- ν is the total number of ions per electrolyte molecule (a dimensionless quantity).

a_{+-}, the activity of the electrolyte, is the term we want to determine. Here $a_\pm = \gamma_\pm m_\pm$ and $\nu = 2$ so by substitution this expression becomes:

$$a_{NaCl} = (\gamma_\pm m_\pm)^2$$

The term $m_\pm = [(m_+)^{\nu^+}(m_-)^{\nu^-}]^{1/\nu}$ can also be substituted to give:

$$a_{NaCl} = \left(\gamma_\pm [(m_+)^{\nu^+}(m_-)^{\nu^-}]^{1/\nu}\right)^2$$

Here $\nu^+ = 1$, $\nu^- = 1$, $\nu = 2$; and $m_+ = m_- = m$ so by substitution this expression becomes:

$$a_{NaCl} = \left(\gamma_\pm [(m)^1(m)^1]^{1/2}\right)^2$$

When simplified this expression becomes:

$$a_{NaCl} = (\gamma_\pm m)^2$$

or:

$$a_{NaCl} = \gamma_\pm^2 m^2$$

where:

- a_{NaCl} is the mean ionic activity (a dimensionless quantity).
- $\gamma_\pm$ is the mean ionic activity coefficient (a dimensionless quantity).
- m is the molality (a dimensionless quantity, obtained by dividing the molality in mol kg^{-1} by 1 mol kg^{-1}).

➔ m represents the molality of the electrolyte. Note that in NaCl the molality of the cations equals the molality of the anions, which in turn equals the molality of the electrolyte, m.

Worked example 6.2B

For a 1:2 electrolyte such as CaCl$_2$ with a molality of m the activity may be determined by considering the number of ions produced upon dissociation, and their molality:

$$CaCl_2 \rightarrow Ca^{2+} + 2\,Cl^-$$
$$a_{+-} = (a_\pm)^\nu$$

Here $a_\pm = \gamma_\pm m_\pm$ and $\nu = 3$ so by substitution this expression becomes:

$$a_{CaCl_2} = (a_\pm)^3 = (\gamma_\pm m_\pm)^3$$

The term $m_\pm = [(m_+)^{\nu^+}(m_-)^{\nu^-}]^{1/\nu}$ can also be substituted to give:

$$a_{CaCl_2} = \left(\gamma_\pm \times [(m_+)^{\nu^+}(m_-)^{\nu^-}]^{1/\nu}\right)^3$$

Here $v^+ = 1$, $v^- = 2$, and $v = 3$; and $m_+ = m$ and $m_- = 2m$ so by substitution this expression becomes:

$$a_{CaCl_2} = \left(\gamma_\pm \times [(m)^1 (2m)^2]^{1/3}\right)^3$$

→ m represents the molality of the electrolyte. Note that in $CaCl_2$ the molality of the cations equals the molality of the electrolyte whereas the molality of the anions is two times the molality of the electrolyte.

When simplified this expression becomes

$$a_{CaCl_2} = \gamma_\pm^3 \times 4m^3$$

> **Question 6.2**
>
> (a) Determine the activity for a 1.00×10^{-2} mol kg^{-1} solution of KCl with a mean activity coefficient of 0.902.
> (b) Determine the activity for a 1.00×10^{-2} mol kg^{-1} solution of $CaCl_2$ with a mean activity coefficient of 0.732.
> (c) An aqueous solution of $CaCl_2$ with a molality of 0.100 mol kg^{-1} has a mean activity coefficient of 0.524 at 298 K. Calculate:
>   i. The mean molality.
>   ii. The mean ionic activity.
>   iii. The activity of the compound.

6.3 Ionic strength

The ionic strength of an electrolyte is defined as:

$$I = \frac{1}{2}\sum_i m_i z_i^2$$

→ Here the symbol Σ represents 'the sum of'. In this case we interpret the overall expression to mean that for each ion in solution we must determine $m_i z_i^2$, and add together the individual values. The ionic strength is then equal to half of this overall value.

where I is the ionic strength (mol kg^{-1}), m_i is the molality of the ion (mol kg^{-1}), and z_i is the charge number of the ion (a dimensionless quantity).

Worked example 6.3A

Calculate the ionic strength of a 0.2 mol kg^{-1} aqueous solution of $CuSO_4$.

Solution

The electrolyte dissociates into its component ions as follows:

$$CuSO_4 \rightarrow Cu^{2+} + SO_4^{2-}$$

The molality of each ion in solution is the same as the molality of the electrolyte, i.e. 0.2 mol kg^{-1}.
The ionic strength is defined as:

$$I = \frac{1}{2}\sum_i m_i z_i^2$$

where:

- I is the ionic strength (mol kg^{-1}).
- m_i is the molality of the ion (mol kg^{-1}).
- z_i is the charge number of the ion (a dimensionless quantity).

$$I = \frac{1}{2}\sum_i m_i z_i^2$$

$$I = \frac{1}{2}\left[(0.2\text{ mol kg}^{-1})\times(+2)^2 + (0.2\text{ mol kg}^{-1})\times(-2)^2\right]$$

$$I = \frac{1}{2}\left[(0.2\text{ mol kg}^{-1})\times(+4) + (0.2\text{ mol kg}^{-1})\times(+4)\right]$$

$$I = \frac{1}{2}\left[(0.8\text{ mol kg}^{-1}) + (0.8\text{ mol kg}^{-1})\right]$$

$$I = \frac{1}{2}\left[1.6\text{ mol kg}^{-1}\right]$$

$$I = 0.8\text{ mol kg}^{-1}$$

➔ Note that for $CuSO_4$ the ionic strength is significantly larger than the molality.

Worked example 6.3B

Calculate the ionic strength of a 0.2 mol kg^{-1} aqueous solution of NaCl.

Solution

The electrolyte dissociates into its component ions as follows:

$$NaCl \rightarrow Na^+ + Cl^-$$

The molality of each ion in solution is the same as the molality of the electrolyte, i.e. 0.2 mol kg^{-1}.
The ionic strength is defined as:

$$I = \frac{1}{2}\sum_i m_i z_i^2$$

$$I = \frac{1}{2}\left[(0.2\text{ mol kg}^{-1})\times(+1)^2 + (0.2\text{ mol kg}^{-1})\times(-1)^2\right]$$

$$I = \frac{1}{2}\left[(0.2\text{ mol kg}^{-1})\times(+1) + (0.2\text{ mol kg}^{-1})\times(+1)\right]$$

$$I = \frac{1}{2}\left[(0.2\text{ mol kg}^{-1}) + (0.2\text{ mol kg}^{-1})\right]$$

$$I = \frac{1}{2}\left[(0.4\text{ mol kg}^{-1})\right]$$

$$I = 0.2\text{ mol kg}^{-1}$$

➔ Note that for NaCl the ionic strength has the same value as the molality.

❓ Question 6.3

(a) Calculate the ionic strength of a 0.2 mol kg^{-1} aqueous solution of Na_2SO_4.
(b) Calculate the ionic strength of an aqueous solution comprising 0.30 mol kg^{-1} Na_2SO_4 and 0.40 mol kg^{-1} $CuSO_4$.

6 ELECTROCHEMISTRY

6.4 Debye–Hückel limiting law

The mean activity coefficient can be determined for an aqueous electrolyte solution at low concentration and 25 °C using the Debye–Hückel limiting law:

$$\log_{10} \gamma_\pm = -0.509 |z_+ z_-| I^{1/2}$$

where:

- $\gamma_\pm$ is the mean activity coefficient of the electrolyte (a dimensionless quantity).
- z_+ and z_- are the charge numbers of the cation and anion respectively (dimensionless quantities).
- I is the ionic strength of the electrolyte (a dimensionless quantity obtained by dividing the ionic strength in mol kg^{-1} by 1 mol kg^{-1}).

→ Note that the number 0.509 relates specifically to an aqueous solution at 298 K, and $|z_+ z_-|$ indicates that it is the modulus of one charge number multiplied by the other that is used. The modulus is simply the numerical value of a parameter irrespective of any sign.

Worked example 6.4A

Estimate the activity coefficient of an aqueous solution of NaCl with an ionic strength of 0.0500 mol kg^{-1}.

Solution

The electrolyte dissociates into its component ions as follows:

$$NaCl \rightarrow Na^+ + Cl^-$$

Here $z_+ = 1$, $z_- = -1$, and $|1 \times -1| = 1$ and:

$$\log_{10} \gamma_\pm = -0.509 |z_+ z_-| I^{1/2}$$
$$\log_{10} \gamma_\pm = -0.509 \times |1 \times 1| \times 0.0500^{1/2}$$
$$\log_{10} \gamma_\pm = -0.509 \times 1 \times 0.0500^{1/2}$$
$$\log_{10} \gamma_\pm = -0.509 \times 1 \times 0.224$$
$$\log_{10} \gamma_\pm = -0.114$$
$$\gamma_\pm = 10^{-0.114}$$
$$\gamma_\pm = 0.769$$

→ Note that the square root is only applied to the final term (the ionic strength) and that if $\log_{10} x = y$ then $x = 10^y$.

Worked example 6.4B

Estimate the activity coefficient of an aqueous solution of ZnCl$_2$ with an ionic strength of 0.0400 mol kg^{-1}.

Solution

The electrolyte dissociates into its component ions as follows:

$$ZnCl_2 \rightarrow Zn^{2+} + 2\,Cl^-$$

Here $z_+ = 2$, $z_- = -1$, and $|2 \times -1| = 2$ and:

$$\log_{10} \gamma_\pm = -0.509 |z_+ z_-| I^{1/2}$$
$$\log_{10} \gamma_\pm = -0.509 \times |2 \times -1| \times 0.0400^{1/2}$$
$$\log_{10} \gamma_\pm = -0.509 \times 2 \times 0.0400^{1/2}$$

$$\log_{10}\gamma_\pm = -0.509 \times 2 \times \sqrt{0.200}$$

$$\log_{10}\gamma_\pm = -0.204$$

$$\gamma_\pm = 10^{-0.204}$$

$$\gamma_\pm = 0.626$$

> Note that the square root is only applied to the final term (the ionic strength) and that if $\log_{10} x = y$ then $x = 10^y$.

> **Question 6.4**
>
> (a) Estimate the activity coefficient of an aqueous solution of Na_2SO_4 with an ionic strength of 0.0900 mol kg^{-1}.
>
> (b) Estimate the activity coefficient of an aqueous solution of KCl with an ionic strength of 0.0200 mol kg^{-1}.

6.5 Conductivity

The number and nature of ions in solution influences conductivity with each electrolyte having a characteristic molar conductivity, Λ_m:

$$\Lambda_m = \frac{\kappa}{c}$$

where:

- Λ_m is the molar conductivity (S m^2 mol^{-1}).
- κ is the conductivity (S m^{-1}).
- c is the concentration of the electrolyte (mol m^{-3}).

> Note that the SI unit for conductivity is siemens (S), where 1 S = 1 Ω^{-1}.

As an electrolyte solution becomes increasingly dilute the molar conductivity, Λ_m, increases until it approaches a limiting value called the limiting molar conductivity, Λ_m^o. The molar conductivity of a strong electrolyte varies with concentration according to **Kohlrausch's law:**

$$\Lambda_m = \Lambda_m^o - \mathcal{K}c^{1/2}$$

where:

- Λ_m is the molar conductivity (S m^2 mol^{-1}).
- Λ_m^o is the limiting molar conductivity (S m^2 mol^{-1}).
- $\mathcal{K}$ is a constant (S m^2 mol^{-1}/(mol m^{-3})$^{1/2}$).
- c is the concentration (mol m^{-3}).

This equation takes the form of the equation for a straight line $y = mx + c$. This form can easily be seen if it is rearranged to give:

$$\Lambda_m = -\mathcal{K}c^{1/2} + \Lambda_m^o$$

The limiting molar conductivity, Λ_m^o (the intercept c) can be obtained by plotting Λ_m (as y) against $c^{1/2}$ (as x). Extrapolation to zero concentration (infinite dilution) will yield the value for Λ_m^o. It has been found that both anions and cations contribute to the limiting molar conductivity:

$$\Lambda_m^o = n_+ \lambda_0^+ + n_- \lambda_0^-$$

> Note a strong electrolyte is one that dissociates completely into its component ions. Molar conductivity decreases as concentration increases due to the effect of long-range electrostatic interactions between charged ions. As concentration increases the plot will begin to deviate from linearity.

where:

- Λ_m^o is the limiting molar conductivity (S m^2 mol^{-1}).
- n_+ is the number of cations per electrolyte molecule.
- λ_0^+ is the limiting molar conductivity of the cation (S m^2 mol^{-1}).

6 ELECTROCHEMISTRY

- n_- is the number of anions per electrolyte molecule.
- λ_0^- is the limiting molar conductivity of the anion (S m^2 mol^{-1}).

The molar conductivity of a weak acid also varies with concentration but in a more complex fashion. This is due to the fact that the degree of dissociation (α) tends to reduce as concentration increases. Hence, Kohlrausch's law only applies to strong electrolytes.

The degree of dissociation (α) of a weak electrolyte is given by the ratio of the molar conductivity to the limiting molar conductivity:

$$\alpha = \frac{\Lambda_m}{\Lambda_m^0}$$

> Note a weak electrolyte is one that partially dissociates into its component ions. Molar conductivity decreases as concentration increases in a non-linear fashion due to the fact that the degree of dissociation varies with concentration: it tends to reduce as concentration increases.

Worked example 6.5A

Find the limiting molar conductivity (Λ_m^0) with units of S cm^2 mol^{-1} for silver nitrate at 298 K from the following data.

c/mol dm^{-3}	0.001	0.01	0.10
Λ_m/S m^2 mol^{-1}	130.5 × 10^{-4}	124.8 × 10^{-4}	109.1 × 10^{-4}

Given that the molar conductivity of the nitrate ion at infinite dilution at 298 K is 71.4 × 10^{-4} S m^2 mol^{-1}, determine the molar conductivity of the silver ion at infinite dilution at 298 K with units of S cm^2 mol^{-1}.

Solution

Plot a graph of molar conductivity against $c^{1/2}$ and extrapolate the straight-line graph to infinite dilution. The SI units of conductivity are S m^2 mol^{-1}. However, the question requests an answer with units of S cm^2 mol^{-1}. This therefore involves multiplying the molar conductivity values provided by 10 000 as there are 10 000 cm^2 in 1 m^2.

c/mol dm^{-3}	0.001	0.01	0.10
$c^{1/2}$/mol$^{1/2}$ dm$^{-3/2}$	0.032	0.10	0.32
Λ_m/S m^2 mol^{-1}	130.5 × 10^{-4}	124.8 × 10^{-4}	109.1 × 10^{-4}
Λ_m/S cm^2 mol^{-1}	130.5	124.8	109.1

A linear fit is applied to the data. (The equation for this fit is shown on the plot in Figure 6.1.) The intercept on the y axis gives the value of the limiting molar conductivity as 132.6 S cm^2 mol^{-1}. Both anions and cations contribute to the limiting molar conductivity:

$$\Lambda_m^0 = n_+\lambda_0^+ + n_-\lambda_0^-$$

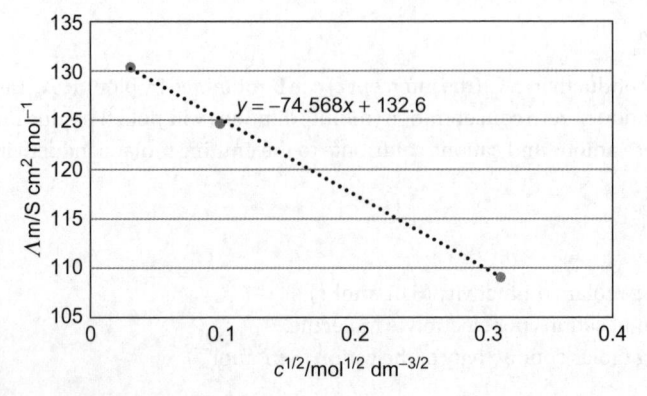

Figure 6.1 Plot of the molar conductivity, Λ_m, against $c^{1/2}$ for aqueous silver nitrate solutions.

where:

- Λ_m^o is the limiting molar conductivity (S m² mol⁻¹).
- n_+ is the number of cations per electrolyte molecule.
- λ_0^+ is the limiting molar conductivity of the cation (S m² mol⁻¹).
- n_- is the number of anions per electrolyte molecule.
- λ_0^- is the limiting molar conductivity of the anion (S m² mol⁻¹).

Hence the limiting molar conductivity of the silver ion is determined by rearranging the expression and replacing known values:

$$\Lambda_m^o = \lambda_0^+ + \lambda_0^-$$

$$\Lambda_m^o - \lambda_0^- = \lambda_0^+$$

$$132.6 \text{ S cm}^2\text{mol}^{-1} - 71.4 \text{ S cm}^2\text{mol}^{-1} = 61.2 \text{ S cm}^2\text{mol}^{-1}$$

The limiting molar conductivity of the silver ion is 61.2 S cm² mol⁻¹.

→ Note the units used here can be expressed differently—for example, in S cm² mol⁻¹—but all terms must have the same units.

→ Note in this case the number of cations per electrolyte molecule is one and the number of anions per electrolyte molecule is also one. So these terms effectively disappear from the expression.

Question 6.5

Find the limiting molar conductivity (Λ_m^o) for potassium chloride at 298 K from the following data:

c/mol dm⁻³	0.001	0.01	0.10
Λ_m/S cm² mol⁻¹	147.0	141.3	129.0

Given that the molar conductivity of the chloride ion at infinite dilution at 298 K is 76.4 S cm² mol⁻¹, determine the molar conductivity of the potassium ion at infinite dilution at 298 K.

Worked example 6.5B

The molar conductivity of sodium ethanoate is found to vary with concentration as follows:

c/mol dm⁻³	0	0.001	0.01	0.10
Λ_m/S cm² mol⁻¹	91.0	88.5	83.8	72.8

Determine the degree of dissociation at each concentration.

Solution

The degree of dissociation (α) of an electrolyte is given by the ratio of the molar conductivity to the limiting molar conductivity:

$$\alpha = \frac{\Lambda_m}{\Lambda_m^o}$$

where:

- α is the degree of dissociation (a dimensionless quantity).
- Λ_m is the molar conductivity (S m² mol⁻¹).
- Λ_m^o is the limiting conductivity (S m² mol⁻¹).

→ Note the units used here can also be expressed in S cm² mol⁻¹ but all terms must have the same units.

At 0.001 mol dm^{-3} $\quad \alpha = \dfrac{\Lambda_m}{\Lambda_m^o} = \dfrac{88.5 \text{ S cm}^2 \text{ mol}^{-1}}{91.0 \text{ S cm}^2 \text{ mol}^{-1}} = 0.973$

At 0.01 mol dm^{-3} $\quad \alpha = \dfrac{\Lambda_m}{\Lambda_m^o} = \dfrac{83.8 \text{ S cm}^2 \text{ mol}^{-1}}{91.0 \text{ S cm}^2 \text{ mol}^{-1}} = 0.921$

At 0.1 mol dm^{-3} $\quad \alpha = \dfrac{\Lambda_m}{\Lambda_m^o} = \dfrac{72.8 \text{ S cm}^2 \text{ mol}^{-1}}{91.0 \text{ S cm}^2 \text{ mol}^{-1}} = 0.800$

> **Question 6.6**
>
> The molar conductivity of ethanoic acid is found to vary with concentration as follows.
>
> | c/mol dm^{-3} | 0 | 0.001 | 0.01 | 0.10 |
> |---|---|---|---|---|
> | Λ_m/S cm^2 mol^{-1} | 390.7 | 48.7 | 16.2 | 5.00 |
>
> Determine the degree of dissociation at each concentration.

6.6 Electrochemical cells

Standard potentials, $E^\ominus$, refer to half-reactions at 298 K in which all ions are at unit activity (1 mol dm^{-3}), all gases are at 1 bar pressure, and all solids are in their most stable form. The following half-reaction, representing the reduction of a proton, has arbitrarily been assigned a potential of zero at 298 K:

$$H^+(aq) + e^- \rightarrow \tfrac{1}{2} H_2(g)$$

The electrochemical series is a list of half-reactions in order of their standard potentials, where reactions that proceed more readily than the reduction of a proton are given a positive value and those which proceed less readily are given a negative value. Standard potentials are thus quoted relative to the standard or normal hydrogen electrode (SHE or NHE respectively) and apply when reactants and products are in their standard states.

The standard Gibbs energy, $\Delta_r G^\ominus$, of the half-reaction is given by:

$$\Delta_r G^\ominus = -nFE^\ominus$$

where:

- $\Delta_r G^\ominus$ is the standard Gibbs energy (J mol^{-1}).
- n is the number of electrons transferred (a dimensionless quantity).
- F is the Faraday constant (96485 C mol^{-1}).
- $E^\ominus$ is the standard cell potential (V).

→ Note that 1 V is 1 J C^{-1} so the units cancel appropriately.

A galvanic cell is one which produces electricity as a consequence of the spontaneous reaction that ensues when two half-cells are coupled together (chemical energy → electrical energy). Some commercial examples of this type of system are the Daniell cell and Leclanché cell.

In an electrochemical cell, oxidations take place at the electrode called the anode and reductions at the electrode called the cathode. When representing a galvanic cell, the oxidation reaction is placed on the left and the reduction reaction on the right; the phase boundaries are represented by a single vertical line and the two half-cells are separated by a double vertical line.

→ Note that the method used here to determine the standard potential of a cell requires an inversion of the sign associated with the anodic reaction.

To calculate the standard potential of a cell formed from any pair of electrodes the following formula can be used:

$$E^\ominus_{\text{Cell}} = E^\ominus_A + E^\ominus_C$$

where:

- $E_{Cell}^{\ominus}$ is the standard cell potential (V).
- $E_A^{\ominus}$ is the standard potential of the anodic oxidation reaction (V).
- $E_C^{\ominus}$ is the standard potential of the cathodic reduction reaction (V).

The standard Gibbs energy, $\Delta_r G^{\ominus}$, of the overall cell reaction is given by the difference of the values of $\Delta_r G^{\ominus}$ for the two half-reactions or by:

$$\Delta_r G^{\ominus} = -nFE_{Cell}^{\ominus}$$

where:

- $\Delta_r G^{\ominus}$ is the standard Gibbs energy (J mol^{-1}).
- n is the number of electrons transferred (a dimensionless quantity).
- F is the Faraday constant (96485 C mol^{-1}).
- $E_{Cell}^{\ominus}$ is the standard cell potential (V).

If the Gibbs energy of the cell reaction or the standard cell potential is known then the equilibrium constant may be determined:

$$\ln K = \frac{-\Delta_r G^{\ominus}}{RT} = \frac{nFE_{Cell}^{\ominus}}{RT}$$

where:

- K is the equilibrium constant (dimensionless).
- R is the gas constant (8.314 J K^{-1} mol^{-1}).
- T is the temperature (K).

For a standard equilibrium reaction such as:

Oxidized species + e$^-$ = Reduced species

the relative activity of the oxidized and reduced species influence the cell potential, so if species are presented at concentrations other than 1 mol dm^{-3} then an equilibrium potential, E_e, can be determined according to the Nernst equation:

$$E_e = E^{\ominus} + \frac{RT}{nF} \ln \frac{a_{\text{oxidized species}}}{a_{\text{reduced species}}}$$

where $a_{\text{oxidized species}}$ is the activity of the oxidized species (a dimensionless quantity) and $a_{\text{reduced species}}$ is the activity of the reduced species (a dimensionless quantity).

In practice, temperature is also a variable but is commonly fixed at a given value. Further, when determining the equilibrium potential, the activity of the species is commonly replaced by concentration to give:

$$E_e = E^{\ominus} + \frac{RT}{nF} \ln \frac{[\text{Oxidized species}]}{[\text{Reduced species}]}$$

➔ If the chemical species are in their standard states then the activity terms are equal to 1 and the logarithm term disappears, making E_e equal to $E^{\ominus}$. Hence $E^{\ominus}$ is referred to as the standard potential.

An electrolytic cell is one in which a non-spontaneous reaction is driven by the application of electrical energy (current or voltage). Some commercial examples of this type of system are electroplating, the chloralkali process (for the production of chlorine and aqueous sodium hydroxide), and the Downs process (for the production of sodium metal).

It is useful to be aware of and understand how to apply Faraday's laws of electrolysis. Faraday's first law relates the mass of substance liberated or electrodeposited at an electrode to the charge passed at the electrode (measured in coulombs, C); Faraday's second law relates the amount in moles of product formed to the amount in moles of electrons transferred.

Worked example 6.6A

Two half-reactions and their standard potentials are:

Half-reaction	$E^\ominus$/V
$Ag^+ + e^- \rightarrow Ag$	+0.80
$Cl_2 + 2\,e^- \rightarrow 2\,Cl^-$	+1.36

Use these standard potentials to determine whether the following equilibrium lies in favour of the reactants or the products and identify the species that undergo oxidation and reduction.

$$2\,Ag(s) + Cl_2(g) \rightleftharpoons 2\,AgCl(s)$$

Solution

The standard potential $E^\ominus\,(Cl_2/Cl^-)$ is more positive than $E^\ominus\,(Ag^+/Ag)$ so the Cl_2 will oxidize the Ag to Ag^+, and will itself be reduced to Cl^-. Hence the reaction shown lies in favour of the products.

Worked example 6.6B

Two half-reactions and their standard potentials are:

Half-reaction	$E^\ominus$/V
$Au^{3+} + 3\,e^- \rightarrow Au$	+1.50
$Cu^{2+} + 2\,e^- \rightarrow Cu$	+0.34

Use these standard potentials to determine whether the following equilibrium lies in favour of the reactants or the products and identify the species that undergo oxidation and reduction.

$$2\,Au(s) + 3\,CuCl_2(aq) \rightleftharpoons 2\,AuCl_3(aq) + 3\,Cu(s)$$

Solution

The standard potential $E^\ominus\,(Au^{3+}/Au)$ is more positive than $E^\ominus\,(Cu^{2+}/Cu)$ so the Au^{3+} will oxidize the Cu to Cu^{2+}, and will itself be reduced to Au. Hence the reaction shown lies in favour of the reactants.

Question 6.7

(a) Two half-reactions and their standard potentials at 298 K are:

Half-reaction	$E^\ominus$/V
$Ni^{2+} + 2\,e^- \rightarrow Ni$	−0.25
$Cu^{2+} + 2\,e^- \rightarrow Cu$	+0.34

Use these standard potentials to determine whether the following equilibrium lies in favour of the reactants or the products and identify the species that undergo oxidation and reduction.

$$Ni(s) + CuCl_2(aq) \rightleftharpoons NiCl_2(aq) + Cu(s)$$

(b) Two half-reactions and their standard potentials at 298 K are:

Half-reaction	$E^\ominus$/V
$Fe^{3+} + e^- \rightarrow Fe^{2+}$	+0.77
$Cu^{2+} + 2e^- \rightarrow Cu$	+0.34

Use these standard potentials to determine whether the following equilibrium lies in favour of the reactants or the products and identify the species that undergo oxidation and reduction.

$$2\,FeCl_3\,(aq) + Cu\,(s) \rightleftharpoons 2\,FeCl_2\,(aq) + CuCl_2\,(aq)$$

(c) Two half-reactions and their standard potentials at 298 K are:

Half-reaction	$E^\ominus$/V
$Zn^{2+} + 2e^- \rightarrow Zn$	−0.76
$Cu^{2+} + 2e^- \rightarrow Cu$	+0.34

Use these standard potentials to determine whether the following equilibrium lies in favour of the reactants or the products and identify the species that undergo oxidation and reduction.

$$ZnCl_2\,(aq) + Cu\,(s) \rightleftharpoons CuCl_2\,(aq) + Zn\,(s)$$

Worked example 6.6C

Consider the following electrochemical cell:

$$Ag\,(s) \,|\, Ag^+\,(aq, 1.00\ mol\ dm^{-3}) \,\|\, Cl^-\,(aq, 1.00\ mol\ dm^{-3}) \,|\, Cl_2\,(g, 1\ bar) \,|\, Pt\,(s)$$

The cell comprises two half-cells. By convention the anode is placed on the left-hand side and the cathode on the right-hand side. The anode is a silver electrode immersed into a 1 mol dm^{-3} aqueous solution of silver ions, and the cathode is a platinum electrode immersed into a 1 mol dm^{-3} aqueous solution of chloride ions with chlorine gas bubbled over it at a pressure of 1 bar. The double vertical line in the middle represents the salt bridge connecting the two half-cells and the single vertical lines represent phase boundaries.

Given the following standard potentials at 298 K:

> Note that the metal electrodes are positioned at the beginning and end of the cell diagram. The anode is on the left-hand side. The cathode is on the right-hand side.

Half reaction	$E^\ominus$/V
$Ag^+ + e^- \rightarrow Ag$	+0.80
$Cl_2 + 2e^- \rightarrow 2\,Cl^-$	+1.36

(a) Determine $E_A^\ominus$ and $E_C^\ominus$ respectively and give the equation that represents the spontaneous cell reaction.
(b) Determine the standard cell potential.
(c) Calculate the standard Gibbs energy change for the reaction.
(d) Calculate the equilibrium constant for the cell reaction.

Solution

(a) The standard potential $E^\ominus\,(Cl_2/Cl^-)$ is more positive than $E^\ominus\,(Ag^+/Ag)$ so the Cl_2 will be reduced to Cl^- and the Ag will be oxidized to Ag^+; hence the silver electrode is the

Table 6.1 Answer summaries for worked example 6.6C part (a).

Half-reaction	Anode or cathode	$E_A^\ominus$/V	$E_C^\ominus$/V
Ag → Ag$^+$ + e$^-$ Note the reaction is inverted	Anode	−0.80 Note the change in sign	
Cl$_2$ + 2 e$^-$ → 2 Cl$^-$	Cathode		+1.36
Spontaneous reaction	2 Ag (s) + Cl$_2$ (g) → 2 AgCl (s)		

anode and the platinum electrode is the cathode; the respective half-reactions are summarized in Table 6.1. $E_A^\ominus$ is given by the standard potential of the silver couple (with the sign inverted), i.e. −0.80 V and $E_C^\ominus$ is the standard potential of the chlorine couple, i.e. +1.36 V. The spontaneous cell reaction leads to the formation of solid silver chloride as shown in Table 6.1.

(b) The cell potential is given by:

$$E_{Cell}^\ominus = E_A^\ominus + E_C^\ominus$$

$$E_{Cell}^\ominus = (-0.80 \text{ V}) + (+1.36 \text{ V})$$

$$E_{Cell}^\ominus = +0.56 \text{ V}$$

(c) The number of electrons transferred is two, so the standard Gibbs energy is given by:

$$\Delta_r G^\ominus = -nFE^\ominus$$

$$\Delta_r G^\ominus = -(2) \times (96485 \text{ C mol}^{-1}) \times (+0.56 \text{ V})$$

$$\Delta_r G^\ominus = -11 \times 10^4 \text{ J mol}^{-1}$$

$$\Delta_r G^\ominus = -110 \text{ kJ mol}^{-1}$$

→ Note that 1 V = 1 J C^{-1} so the units of Gibbs energy are J mol^{-1} or kJ mol^{-1}. As expected for a spontaneous reaction, the value for the Gibbs energy change is negative. This value is −110 kJ mol^{-1} per two moles of silver or per one mole of chlorine.

(d) The equilibrium constant is given by:

$$\ln K = \frac{nFE_{Cell}^\ominus}{RT}$$

$$\ln K = \frac{(2) \times (96\,485 \text{ C mol}^{-1}) \times (+0.56 \text{ V})}{(8.314 \text{ J K}^{-1}\text{mol}^{-1}) \times (298 \text{ K})}$$

$$K = e^{\frac{(2) \times (96\,485 \text{ C mol}^{-1}) \times (+0.56 \text{ V})}{(8.314 \text{ J K}^{-1}\text{mol}^{-1}) \times (298 \text{ K})}}$$

$$K = 8.8 \times 10^{18}$$

→ Note that 1 V = 1 J C^{-1} hence the units all cancel. Also note that the value for K is rather large, as expected for a spontaneous reaction.

Worked example 6.6D

A half cell containing a copper electrode immersed in a 1 mol dm^{-3} aqueous solution of CuSO$_4$ is connected via a salt bridge to a half cell containing a gold electrode immersed in a 1 mol dm^{-3} aqueous solution of AuCl$_3$.

Given the following standard potentials at 298 K:

Half-reaction	$E^\ominus$/V
Au^{3+} + 3 e$^-$ → Au	+1.50
Cu^{2+} + 2 e$^-$ → Cu	+0.34

(a) Identify the cathode and the anode and give the cell diagram that represents the overall electrochemical cell.

6.6 ELECTROCHEMICAL CELLS

(b) Determine $E_A^\ominus$ and $E_C^\ominus$ respectively and give the equation that represents the spontaneous cell reaction.

(c) Determine the standard cell potential.

(d) Calculate the standard Gibbs energy change for the reaction.

(e) Calculate the equilibrium constant for the cell reaction.

Solution

(a) The standard potential $E^\ominus$ (Au^{3+}/Au) is more positive than $E^\ominus$ (Cu^{2+}/Cu) so the Au^{3+} will be reduced to Au and the Cu will be oxidized to Cu^{2+}. Hence the copper electrode is the anode and the gold electrode is the cathode. By convention the anode is placed on the left-hand side and the cathode on the right-hand side so the cell diagram is given by:

Cu (s) | Cu^{2+} (aq, 1.00 mol dm^{-3}) ‖ Au^{3+} (aq, 1.00 mol dm^{-3}) | Au (s)

(b) The respective half reactions are summarized in Table 6.2. $E_A^\ominus$ is given by the standard potential of the copper couple (with the sign inverted), i.e. −0.34 V and $E_C^\ominus$ is the standard potential of the gold couple, i.e. +1.50 V. The spontaneous cell reaction leads to the formation of solid gold and the dissolution of copper as shown in the table below.

(c) The cell potential is given by:

$E_{Cell}^\ominus = E_A^\ominus + E_C^\ominus$

$E_{Cell}^\ominus = (-0.34\ V) + (+1.50\ V)$

$E_{Cell}^\ominus = +1.16\ V$

(d) The number of electrons transferred is six, so the standard Gibbs energy is given by

$\Delta_r G^\ominus = -nFE^\ominus$

$\Delta_r G^\ominus = -(6) \times (96485\ C\ mol^{-1}) \times (+1.16\ V)$

$\Delta_r G^\ominus = -672 \times 10^3\ C\ V\ mol^{-1} = -672\ kJ\ mol^{-1}$

→ Note that 1 V = 1 J C^{-1} so the units of Gibbs energy are J mol^{-1} or kJ mol^{-1}. The value is negative indicating that the reaction is spontaneous. Also note that this value is −672 kJ mol^{-1} per two moles of Au^{3+} or per three moles of copper.

(e) The equilibrium constant is given by:

$\ln K = \dfrac{nFE_{Cell}^\ominus}{RT}$

$\ln K = \dfrac{(6) \times (96485\ C\ mol^{-1}) \times (+1.16\ V)}{(8.314\ J\ K^{-1}\ mol^{-1}) \times (298\ K)}$

$K = e^{\frac{(6) \times (96485\ C\ mol^{-1}) \times (+1.16\ V)}{(8.314\ J\ K^{-1}\ mol^{-1}) \times (298\ K)}}$

$K = 5.2 \times 10^{117}$

→ Note this value for K is very large. Most calculators will be unable to determine an answer and will return an error message at the final step.

→ Note that 1 V = 1 J C^{-1} hence the units all cancel.

Table 6.2 Answer summaries for worked example 6.6D part (b).

Half-reaction	Anode or cathode	$E_A^\ominus$/V	$E_C^\ominus$/V
3 Cu → 3 Cu^{2+} + 6 e$^-$ Note the reaction is inverted	Anode	−0.34 Note the change in sign	
2 Au^{3+} + 6 e$^-$ → 2 Au	Cathode		+1.50
Spontaneous reaction	2 Au^{3+}(aq) + 3 Cu (s) → 2 Au (s) + 3 Cu^{2+}(aq)		

Question 6.8

(a) Consider the following electrochemical cell:

Zn (s) | Zn^{2+} (aq, 1.00 mol dm^{-3}) || Cu^{2+} (aq, 1.00 mol dm^{-3}) | Cu (s)

Given the following standard potentials at 298 K:

Half-reaction	$E^\ominus$/V
Zn^{2+} + 2 e$^-$ → Zn	−0.76
Cu^{2+} + 2 e$^-$ → Cu	+0.34

i. Determine $E_A^\ominus$ and $E_C^\ominus$ respectively and give the equation that represents the spontaneous cell reaction.
ii. Determine the standard cell potential.
iii. Calculate the standard Gibbs energy change for the reaction.
iv. Calculate the equilibrium constant for the cell reaction.

(b) A half-cell containing a nickel electrode immersed in a 1 mol dm^{-3} aqueous solution of NiSO$_4$ is connected via a salt bridge to a half-cell containing a copper electrode immersed in a 1 mol dm^{-3} aqueous solution of CuSO$_4$.

Given the following standard potentials at 298 K:

Half-reaction	$E^\ominus$/V
Ni^{2+} + 2 e$^-$ → Ni	−0.25
Cu^{2+} + 2 e$^-$ → Cu	+0.34

i. Identify the cathode and the anode and give the cell diagram that represents the overall electrochemical cell.
ii. Determine $E_A^\ominus$ and $E_C^\ominus$ respectively and give the equation that represents the spontaneous cell reaction.
iii. Determine the standard cell potential.
iv. Calculate the standard Gibbs energy change for the reaction.
v. Calculate the equilibrium constant for the cell reaction.

(c) Consider the following electrochemical cell:

Cu (s) | Cu^{2+} (aq, 1.00 mol dm^{-3}) || Fe^{3+} (aq, 1.00 mol dm^{-3}) | Fe^{2+} (aq, 1.00 mol dm^{-3}) | Pt (s)

Two half reactions and their standard potentials are given at 298 K:

Half-reaction	$E^\ominus$/V
Fe^{3+} + e$^-$ → Fe^{2+}	+0.77
Cu^{2+} + 2 e$^-$ → Cu	+0.34

i. Determine $E_A^\ominus$ and $E_C^\ominus$ respectively and give the equation that represents the spontaneous cell reaction.
ii. Determine the standard cell potential.
iii. Calculate the standard Gibbs energy change for the reaction.
iv. Calculate the equilibrium constant for the cell reaction.

Worked example 6.6E

In a solution of 0.03 mol dm^{-3} oxidized species and 0.03 mol dm^{-3} reduced species, given that the standard potential is +0.65 V, $n = 2$, and $T = 25$ °C, what is the equilibrium potential?

Solution

$$E_e = E^\ominus + \frac{RT}{nF} \ln \frac{[\text{Oxidized species}]}{[\text{Reduced species}]}$$

$$E_e = (+0.65 \text{ V}) + \left(\frac{(8.314 \text{ J K}^{-1}\text{mol}^{-1}) \times (298 \text{ K})}{2 \times (96485 \text{ C mol}^{-1})} \times \ln \frac{[0.03]}{[0.03]} \right)$$

$$E_e = +0.65 \text{ V}$$

→ Note that 1 V = 1 J C^{-1} hence the units of E_e are the same as the units for $E^\ominus$. Also note that when the concentrations of both species are equal then E_e has the same value as $E^\ominus$.

Worked example 6.6F

In a solution of 0.06 mol dm^{-3} oxidized species and 0.03 mol dm^{-3} reduced species, given that the standard potential is +0.55 V, $n = 2$, and $T = 25$ °C, what is the equilibrium potential?

Solution

$$E_e = E^\ominus + \frac{RT}{nF} \ln \frac{[\text{Oxidized species}]}{[\text{Reduced species}]}$$

$$E_e = (+0.55 \text{ V}) + \left(\frac{(8.314 \text{ J K}^{-1}\text{mol}^{-1}) \times (298 \text{ K})}{2 \times (96485 \text{ C mol}^{-1})} \times \ln \frac{[0.06]}{[0.03]} \right)$$

$$E_e = +0.56 \text{ V}$$

→ Note that 1 V = 1 J C^{-1} hence the units of E_e are the same as the units for $E^\ominus$.

❓ Question 6.9

(a) In a solution of 0.10 mol dm^{-3} oxidized species and 0.01 mol dm^{-3} reduced species, given that the standard potential is +0.42 V, $n = 2$, and $T = 25$ °C, what is the equilibrium potential?

(b) What is the equilibrium potential of a half-cell composed of a copper wire immersed into a 0.030 mol dm^{-3} aqueous solution of Cu^{2+} ions at 25 °C?

(c) What is the equilibrium potential of a half-cell composed of a nickel wire immersed into a 0.050 mol dm^{-3} aqueous solution of Ni^{2+} ions at 35 °C?

(d) What is the equilibrium potential of a half-cell composed of a silver wire immersed into a 0.080 mol dm^{-3} aqueous solution of Ag$^+$ ions at 45 °C?

Worked example 6.6G

During the electrolysis of an aqueous copper (II) sulfate solution a current of 100 mA was passed for 75 min. What mass of copper metal was deposited at the cathode?

Solution

The reaction at the cathode is:

$$Cu^{2+}(aq) + 2\ e^- \rightarrow Cu(s)$$

So deposition of one mole of copper requires two moles of electrons corresponding to a charge of:

$$Q = (2\ \text{mol}) \times (96485\ \text{C mol}^{-1})$$

$$Q = 192970\ \text{C}$$

Compare this value with the charge passed during the electrolysis, as determined from the information provided using:

$$Q = It$$

where Q is the electrical charge (C), I is the current (A), and t is the time (s).
The current should be converted to SI units: 100 mA = $((1 \times 10^{-3}\ \text{A mA}^{-1}) \times 100\ \text{mA}) = 0.100\ \text{A}$.
The time should be converted to SI units: $(75\ \text{min}) \times (60\ \text{s min}^{-1}) = 4500\ \text{s}$.

$$Q = It$$

$$Q = (0.100\ \text{A}) \times (4500\ \text{s})$$

Note that $1\ \text{A} = 1\ \text{C s}^{-1}$ hence:

$$Q = (0.100\ \text{C s}^{-1}) \times (4500\ \text{s})$$

$$Q = 450\ \text{C}$$

The actual amount in moles of copper deposited during electrolysis is given by:

$$\text{Amount in moles deposited} = \frac{(450\ \text{C})}{(192970\ \text{C mol}^{-1})}$$

$$\text{Amount in moles deposited} = 2.33 \times 10^{-3}\ \text{mol}$$

The molar mass of copper is 63.54 g mol^{-1} so the actual mass of copper deposited during electrolysis is given by:

$$\text{Mass deposited} = (2.33 \times 10^{-3}\ \text{mol}) \times (63.54\ \text{g mol}^{-1})$$

$$\text{Mass deposited} = 0.148\ \text{g}$$

Worked example 6.6H

During the electrolysis of an aqueous Ag$^+$ solution, Ag is deposited at the cathode. For how long would a current of 10 mA need to be passed in order to produce 33.5×10^{-3} g of Ag at the cathode?

Solution

The molar mass of silver is 107.87 g mol^{-1} so the amount in moles of silver deposited during electrolysis is given by:

Amount in moles deposited $= \dfrac{(33.5 \times 10^{-3}\text{ g})}{(107.87\text{ g mol}^{-1})}$

Amount in moles deposited $= 311 \times 10^{-6}$ mol

The reaction at the cathode is:

$$Ag^+(aq) + e^- \rightarrow Ag(s)$$

So deposition of 311×10^{-6} mol of silver requires 311×10^{-6} mol of electrons corresponding to a charge of:

$Q = (311 \times 10^{-6}\text{ mol}) \times (96485\text{ C mol}^{-1})$

$Q = 30.0$ C

Remember that:

$Q = It$

where Q is the electrical charge (C), I is the current (A), and t is the time (s).

So this can be rearranged to make time the subject:

$t = \dfrac{Q}{I}$

The current should be converted to SI units: 10 mA $= ((1 \times 10^{-3}\text{ A mA}^{-1}) \times 10\text{ mA}) = 0.010$ A.

$t = \dfrac{(30.0\text{ C})}{(0.010\text{ A})}$

Note that $1\text{ A} = 1\text{ C s}^{-1}$ hence:

$t = \dfrac{(30.0\text{ C})}{(0.010\text{ C s}^{-1})}$

$t = 3000$ s

The actual time taken to deposit 33.5×10^{-3} g of silver is 3000 s.

❓ Question 6.10

(a) During the electrolysis of an aqueous Pb^{2+} solution a current of 25 mA was passed for 30 min. What mass of lead metal was deposited at the cathode?

(b) During the electrolysis of an aqueous Sn^{2+} solution, Sn is deposited at the cathode. For how long would a current of 34 mA need to be passed in order to produce 0.132 g of Sn at the cathode?

Turn to the Synoptic questions section on page 175 to attempt questions that encourage you to draw on concepts and problem-solving strategies from several topics within a given chapter to come to a final answer.

Final answers to numerical questions appear at the end of the book, and full worked solutions appear on the book's website, where you can also find a set of bonus questions for each chapter. Go to www.oxfordtextbooks.co.uk/orc/chemworkbooks/.

Synoptic questions

The following questions have been written to help you draw on the various topics covered within each chapter in an overarching way. As such, these questions may require you to use concepts and problem-solving strategies from several topics to come to a final answer. Final answers to numerical questions appear at the end of the book, and full worked solutions appear on the book's website, where you can also find a set of bonus questions for each chapter. Go to www.oxfordtextbooks.co.uk/orc/chemworkbooks/.

Chapter 1 Fundamentals

Question S1.1

A sample of impure copper metal was analysed for the percentage purity of copper. A 5.014 g sample of the metal was dissolved in concentrated nitric acid and the solution made up to 500 cm^3 in a graduated flask. Portions (25.00 cm^3) of the solution were measured and added to a conical flask. KI solution (20 cm^3) of approx. 0.1 M was added to the copper solution and the iodine released was titrated against sodium thiosulfate solution of molarity 0.1500 M. An average of 24.35 cm^3 Na$_2$S$_2$O$_3$ solution was required to completely react with all the iodine released.

Write half equations for the reduction of Cu^{2+} by KI and the reduction of the iodine released by sodium thiosulfate. Combine the equations to obtain the number of moles of Cu^{2+} equivalent to one mole of thiosulfate and hence determine the percentage purity of the metal.

Question S1.2

Calculate the total mass of gas in a typical student room of dimensions 3 m × 4 m × 3 m assuming the gas is 80% nitrogen and 20% oxygen and 1 mole of gas occupies 24 dm^3 under normal conditions of temperature and pressure.

Question S1.3

The molecule commercially known as desflurane is used as an anaesthetic gas. The gas contains 21.4% carbon, 68% fluorine, 9.5% oxygen, and 1.1% hydrogen by mass.

(a) Determine the empirical formula of the gas. The mass spectrum of the compound shows the presence of the parent ion at 168 g mol^{-1}. Hence determine the molecular formula of the compound.

(b) The gas is normally administered with nitrous oxide and oxygen in the molar ratio:
Nitrous oxide:oxygen:desflurane = 10:5:1. Calculate the partial pressure of each gas in an anaesthetic mixture administered at 1 atmosphere pressure.

Question S1.4

Each square centimetre of the earth's surface supports a column of air with mass 1.0 kg.

(a) Calculate the mass of the atmosphere.

(b) Assuming that the average molar mass of gases constituting the atmosphere is 28.8 g mol^{-1} and that the average temperature is 20 °C, calculate the number of moles of gas in the atmosphere.

(The area of a sphere of radius R is $4\pi R^2$ and the radius of the earth is 6400 km.)

Question S1.5

An impure sample of WCl$_6$ was prepared and 0.0216 g of the sample was analysed by titration with AgNO$_3$ solution. AgNO$_3$ solution (23.00 cm^3) of concentration 0.0105 M was required to completely react with the chloride ions from WCl$_6$. Calculate the percentage purity of the WCl$_6$ sample.

Chapter 2 Thermodynamics

Question S2.1

(a) Define an enthalpy change in terms of changes in internal energy and volume.

(b) Give a statement of Hess's Law.

(c) From the following data, determine $\Delta_f H^\ominus$ for diborane, B$_2$H$_6$ (g), at 298 K:

$B_2H_6 (g) + 3 O_2 (g) \rightarrow B_2O_3 (s) + 3 H_2O (g)$ $\quad \Delta_r H^\ominus = -1941$ kJ mol^{-1}

$2 B (s) + \frac{3}{2} O_2 (g) \rightarrow B_2O_3 (s)$ $\quad \Delta_r H^\ominus = -2368$ kJ mol^{-1}

$H_2 (g) + \frac{1}{2} O_2 (g) \rightarrow H_2O (g)$ $\quad \Delta_r H^\ominus = -241.8$ kJ mol^{-1}

Question S2.2

Nitric acid (HNO$_3$) reacts with hydrazine (N$_2$H$_4$) according to the reaction equation:

$4 HNO_3 (l) + 5 N_2H_4 (l) \rightarrow 7 N_2 (g) + 12 H_2O (l)$

$\Delta_f H^\ominus(HNO_3 (l)) = -174.1$ kJ mol^{-1}

$\Delta_f H^\ominus(N_2H_4 (l)) = +50.63$ kJ mol^{-1}

$\Delta_f H^\ominus(H_2O (l)) = -285.8$ kJ mol^{-1}

Using the data given above, calculate the standard molar enthalpy of this reaction, $\Delta_r H^\ominus$.

Question S2.3

The reaction $2 SO_2 (g) + O_2 (g) \rightleftharpoons 2 SO_3 (g)$ has an equilibrium constant $K = 3.0 \times 10^4$ at a temperature of 700 K. Calculate the Gibbs energy change, $\Delta_r G^\ominus$, for this reaction and state clearly whether the equilibrium favours reactants or products at this temperature.

Question S2.4

(a) Calculate the work that must be done against the atmosphere during the combustion of 1 mole of benzene at 25 °C and 1 atmosphere pressure.

(b) Given that the enthalpy of combustion of benzene is -3273 kJ mol^{-1} calculate the change in internal energy in the system.

Question S2.5

Use the bond enthalpies given in Question 2.13 (see Chapter 2, section 2.5) to determine the additional energy released from the complete combustion of one mole of propane compared to one mole of butane. How does this compare to the experimental values for enthalpies of combustion of the two hydrocarbons ($\Delta_c H^\circ(C_3H_8(g)) = 2220$ kJ mol^{-1}, $\Delta_c H^\circ(C_4H_{10}(g)) = 2877$ kJ mol^{-1})?

Question S2.6

Calculate the temperature at which it becomes thermodynamically possible for the following reaction to occur: $2\,Fe_2O_3(s) + 3\,Cs(s) \rightleftharpoons 4\,Fe(s) + 3\,CO_2(g)$

$\Delta_f H^\circ (CO_2(g)) = -393.5$ kJ mol^{-1}

$\Delta_f H^\circ (Fe_2O_3(s)) = -824.2$ kJ mol^{-1}

$_f S^\circ(Fe(s)) = 27.3$ J K^{-1} mol^{-1}

$_f S^\circ(CO_2(g)) = 213.7$ J K^{-1} mol^{-1}

$_f S^\circ(Fe_2O_3(s)) = 87.4$ J K^{-1} mol^{-1}

$_f S^\circ(C(s)) = 5.7$ J K^{-1} mol^{-1}

Chapter 3 Chemical equilibrium

Question S3.1

Given that the standard Gibbs energies of formation of NO$_2$ and N$_2$O$_4$ are 51.8 kJ mol^{-1} and 98.3 kJ mol^{-1} respectively, calculate $\Delta_r G^\circ$, K_p, and K_c at one atmosphere pressure and 25 °C for the following reaction: $N_2O_4(g) \rightleftharpoons 2\,NO_2(g)$.

Question S3.2

Consider the reaction below and determine the percentage dissociation of N$_2$O$_4$ at 127 °C.

$N_2O_4(g) \rightleftharpoons 2\,NO_2(g) \quad K_p = 47.9$ at 400 K

Question S3.3

In a solution containing 0.15 mol dm^{-3} Cd^{2+} what concentration of sulfide ions will result in the precipitation of CdS? (Note that K_{sp} for the salt is 1.6×10^{-28} mol^2 dm^{-6}.)

Question S3.4

Consider the following equilibrium:

$2\,A(g) + 2\,B(s) \rightleftharpoons 4\,C(g) \quad K_p = 3.00 \times 10^3$ atm^2 at 600 K

(a) Give an expression for K_p for the above reaction.

(b) The standard reaction enthalpy, $\Delta_r H^\circ$, is -91.0 kJ mol^{-1}. Estimate the value of K_p at 675 K.

(c) Determine K_c for the equilibrium at 600 K.

(d) What is the value of K_p for the reverse reaction at 600 K?

Question S3.5

Calculate the hydrated proton ion concentration and the sulfide ion concentration in a 0.15 mol dm^{-3} aqueous solution of H$_2$S. The dissociation constant for the first ionization, pK_1, has a value of 7.05 and that of the second ionization, pK_2, has a value of 12.92.

Chapter 4 Phase Equilibrium

Question S4.1

Two partially miscible liquids are mixed thoroughly. Upon standing, the solution separates into two saturated layers. Determine the number of components, C, the number of phases, P, and the number of degrees of freedom, F, in the system.

Question S4.2

Figure S1 represents a phase diagram. Identify the phase(s) present at points a, b, c, and d.

Question S4.3

Determine the enthalpy of vaporization, $\Delta_{vap}H$, for benzene assuming there are no specific molecular interactions, i.e. no hydrogen bonding. Benzene has a normal boiling point, $T_b = 353$ K.

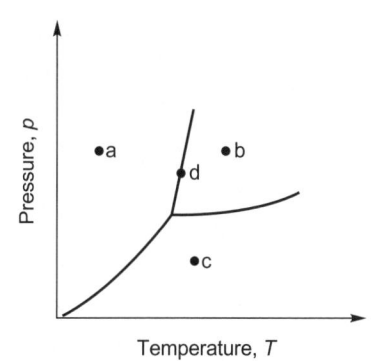

Figure S1 Phase diagram for Question S4.2.

Question S4.4

Calculate the entropy change when 18 g of water vaporizes at 100.00 °C. (The standard enthalpy change of vaporization is +40.7 kJ mol^{-1}.)

Question S4.5

0.75 g of a protein in 100 cm^3 of water exerts an osmotic pressure of 232 Pa at 25 °C. Assume the solution is ideal and determine the molecular mass of the protein.

Chapter 5 Reaction kinetics

Question S5.1

The radioactive iodine isotope ^{131}I is commonly used to treat an overactive thyroid. The decay of ^{131}I follows first-order kinetics and has a half-life of 8.02 days.

(a) Write a rate equation for the decay of ^{131}I.

(b) What is meant by a **half-life** and how is it related to the first-order rate constant? Calculate the first-order rate constant for this decay.

(c) A single dose of therapeutic Na^{131}I solution was produced with an activity of 4.5×10^9 decays per second. The minimum activity that can be administered to a patient is 2.6×10^8 decays per second. What is the maximum time after production that this solution can be used?

Question S5.2

The catalytic decomposition of ammonia on hot tungsten is a zero-order reaction when the partial pressure of ammonia is sufficiently high.

In an experiment, the zero-order rate constant for ammonia decomposition was found to be 1.43×10^{-2} kPa s^{-1}. The initial partial pressure of ammonia was 30 kPa. Calculate the partial pressure of ammonia after 10 minutes and determine how long it will take for all the ammonia to disappear.

Question S5.3

The equation for the hydrolysis of bromoethane in alkaline solution is given by:

$$C_2H_5Br\,(aq) + OH^-\,(aq) \rightarrow C_2H_5OH\,(aq) + Br^-\,(aq)$$

The rate constant, k, for the reaction was measured at a series of temperatures; the results are given below. Use the results to find a value for the activation energy for the reaction.

T/K	$k \times 10^4$/mol^{-1} s^{-1}
298	0.85
301	1.30
304	1.90
307	2.50
310	3.70
313	5.10
316	7.00
319	9.60

Question S5.4

The rate constant for the first-order decomposition of N$_2$O$_5$ which proceeds according to the reaction:

$$2\,N_2O_5\,(g) \rightarrow 4NO_2\,(g) + O_2\,(g)$$

is $k = 3.35 \times 10^{-5}$ s^{-1} at 298 K.

(a) Calculate the half-life for the decomposition of N$_2$O$_5$ in this reaction.

(b) If the initial pressure of N$_2$O$_5$ is 500 Torr calculate the pressure of N$_2$O$_5$ after 10 min.

Question S5.5

If a reaction obeys second-order kinetics show that the time taken for three quarters of the original concentration to disappear is three times the half-life.

Question S5.6

A proposed mechanism for the reaction (1) is given by:

$$2\text{ NO} \underset{k_{-1}}{\overset{k_1}{\rightleftharpoons}} N_2O_2$$

$$N_2O_2 + H_2 \xrightarrow{k_2} N_2O + H_2O$$

Reaction (1): $2\text{ NO (g)} + H_2\text{ (g)} \rightarrow N_2O\text{ (g)} + H_2O\text{ (g)}$

Show that the rate law can be expressed by Rate $= \dfrac{d[N_2O]}{dt} = k_{obs}[NO]^2[H_2]$ and state what assumptions you have made in deriving this.

Chapter 6 Electrochemistry

Question S6.1

An aqueous solution of $MgCl_2$ with a molality of 0.110 mol kg^{-1} is prepared; estimate its activity coefficient and its activity.

Question S6.2

At 298 K the molar conductivity of aqueous lithium chloride solutions was found to vary with concentration as shown in this data table:

c/mol dm^{-3}	0.001	0.01	0.10
Λ_m/S m^2 mol^{-1}	112.4 × 10^{-4}	107.3 × 10^{-4}	95 × 10^{-4}

Using this data:

(a) Find the limiting molar conductivity (Λ_m^o) for lithium chloride.
(b) Determine Kohlrausch's constant for this system.
(c) Given that the molar conductivity of the chloride ion at infinite dilution at 298 K is 76.4 × 10^{-4} S m^2 mol^{-1}, determine the molar conductivity of the lithium ion at infinite dilution.

Question S6.3

The standard potential of the Fe^{3+}/Fe^{2+} redox couple is +0.77 V; in a solution containing 0.03 mol dm^{-3} Fe^{3+} and 0.05 mol dm^{-3} Fe^{2+}, what is the equilibrium potential? Assume that $n = 1$ and $T = 25\,°C$.

Question S6.4

During the electrolysis of an aqueous nickel sulfate solution a current of 76.0 mA was passed for 85.0 min. What mass of nickel metal was deposited at the cathode?

Answers

Final answers to questions posed in the text (where they can be given) are presented here. You can find **full solutions** to every question featured in the book in the *Workbooks in Chemistry* Online Resource Centre. Go to www.oxfordtextbooks.co.uk/orc/chemworkbooks/.

CHAPTER 1

1.1 $kg\,m^{-3}$
1.2 $kg\,m\,s^{-2}$ = newton (N)
1.3 $kg\,m^{-1}\,s^{-2}$ = pascal (Pa)
1.4. (a) $1 \times 10^{-1}\,m^3$
   (b) $1 \times 10^4\,mol\,dm^{-3}$
   (c) $1 \times 10^{-2}\,mol\,cm^{-3}$
   (d) $1 \times 10^2\,mol\,dm^{-3}$
1.5 (a) 2.002 m
   (b) 3.5 m
   (c) $2.950 \times 10^{-4}\,m$
1.6 2.89 kJ
1.7 $2.4 \times 10^2\,dm^3$
1.8 i. 3.01×10^{23} atoms
   ii. 6.02×10^{23} atoms
1.9 3.011×10^{24} ions
1.10 24.3
1.11 90.09 g
1.12 0.154 mol
1.13 Br^-
1.14 C_5H_7N (empirical formula) $C_{10}H_{14}N_2$ (molecular formula)
1.15 (a) Cr_2O_3
    (b) Cr_2O_3
1.16 15.9%
1.17 Cl
1.18 0.122 g
1.19 $0.0107\,mol\,dm^{-3}$
1.20 $0.12\,mol\,dm^{-3}$
1.21 $50\,cm^3$
1.22 88%
1.23 (a) $100\,cm^3$
    (b) 4.61 g
1.24 68%
1.25 $6.5\,kg\,m^{-3}$
1.26 $132\,g\,mol^{-1}$
1.27 $x_{CO_2} = 0.937$ $p_{CO_2} = 562.2\,Pa$
    $x_{N_2} = 0.0423$ $p_{N_2} = 25.38\,Pa$
    $x_{Ar} = 0.0207$ $p_{Ar} = 12.42\,Pa$
1.28 3.69 atm
1.29 0.266 (CO_2) 0.734 (CH_4)

CHAPTER 2

2.1 (a) open
   (b) closed
   (c) closed
   (d) theoretically closed, in practice open
2.2 (a) state
   (b) state
   (c) state
   (d) path
2.3 (a) intensive
   (b) extensive
   (c) extensive
   (d) extensive
2.4 +2.43 kJ
2.5 −1.1 kJ
2.6 (a) −405 J
   (b) −340 J
2.7 −696 J and −912 J
2.8 −2.5 kJ
2.9 −247.9 kJ
2.10 −5316 kJ
2.11 +15.4 kJ
2.12 (b) $+697\,kJ\,mol^{-1}$
    (c) $-147\,kJ\,mol^{-1}$
2.13 $-347\,kJ\,mol^{-1}$
2.14 $-2054\,kJ\,mol^{-1}$
2.15 $-582\,kJ\,mol^{-1}$ and $-1154\,kJ\,mol^{-1}$
2.16 553 kJ
2.17 $0.43\,J\,g^{-1}\,°C^{-1}$
2.18 1382 kJ
2.19 $2000\,kJ\,mol^{-1}$
2.20 (a) increase
    (b) increase
    (c) decrease
    (d) increase
2.21 $87.2\,J\,K^{-1}\,mol^{-1}$
2.22 $4.37\,J\,K^{-1}\,mol^{-1}$
2.23 $-285\,J\,K^{-1}$
2.24 $-131\,J\,K^{-1}$
2.25 (a) $-166\,kJ\,mol^{-1}$
    (b) Increase, +ve
    (c) $+236\,J\,K^{-1}\,mol^{-1}$
2.26 $\Delta S_{tot} = -92.7\,J\,K^{-1}$ not spontaneous
2.27 $+34.7\,kJ\,mol^{-1}$ not spontaneous
2.28 $-856.1\,kJ\,mol^{-1}$ spontaneous
2.29 $-140\,kJ\,mol^{-1}$
2.30 2.0×10^{-5}
2.31 (a) i. $-29\,kJ\,mol^{-1}$,
      ii. 1.21×10^5
    (b) 976 K

CHAPTER 3

3.2 (a) $361\,mol\,dm^{-3}$
   (b) $43.9 \times 10^3\,mol^2\,dm^{-6}$
   (c) $9.8 \times 10^{25}\,mol^{-1}\,dm^3$
   (d) 54
3.3 (a) $0.0245\,atm^2$
   (b) $2.5 \times 10^{-25}\,atm$
   (c) 0.019
3.4 (a) 0.269
   (b) $26.8\,kJ\,mol^{-1}$
3.5 (a) 1.1×10^{-2}
   (b) 3.3

3.6   24 mol dm^{-3}

3.7   [H$_2$] = 0.37 mol dm^{-3}; [I$_2$] = 1.37 mol dm^{-3}; [HI] = 5.25 mol dm^{-3}

3.8   Equilibrium partial pressure of SO$_3$ = 1.50 atm; Equilibrium partial pressure of SO$_2$ = 3.43 × 10^{-4} atm; Equilibrium partial pressure of O$_2$ = 1.72 × 10^{-4} atm

3.9   (a) 8.8 × 10^{-7} mol dm^{-3}
      (b) 1.6 × 10^{-5} mol dm^{-3}
      (c) 3.2 × 10^{-28} mol^2 dm^{-6}

3.10 (a) −10.0
      (b) 6.35

3.11 (a) 1.7 × 10^{-5} mol dm^{-3}
      (b) 1 × 10^{-9} mol dm^{-3}

3.12 (a) 1.66 and 12.34
      (b) 12.68 and 1.32

3.13 2.58

CHAPTER 4

4.1   (a) 12 × 10^6 Pa
      (b) 0.0448 m^3

4.2   (a) $F = 2$
      (b) $F = 2$
      (c) $F = 2$
      (d) $F = 1$
      (e) $F = 0$

4.3   (a) 26 kJ mol^{-1}
      (b) 27 kJ mol^{-1}

4.4   (a) 310 K
      (b) 420 K

4.5   (a) 50.05 °C
      (b) 78.57 °C

4.6   (a) 349 K
      (b) 422 K

4.7   (a) 206.6 torr
      (b) p_2 = 0.1294 atm

4.8   (a) 17.1 kJ mol^{-1}
      (b) 40.3 kJ mol^{-1}

4.9   (a) −0.13 °C
      (b) 4.5 °C

4.10 (a) 81.0 °C
      (b) 100.8 °C

4.11 (a) 5.3 atm
      (b) 3 × 10^{-3} atm

4.12 (a) 2.1 × 10^3 g mol^{-1}
      (b) 33 g mol^{-1}

4.13 (a) 6 × 10^3 g mol^{-1}
      (b) 856 g mol^{-1}

4.14 (a) 78.2 torr
      (b) 74.6 kPa

4.15 (a) The mole fraction of benzene in the vapour is 0.21 and the mole fraction of methylbenzene in the vapour is 0.79
      (b) The mole fraction of liquid A in the vapour is 0.327 and the mole fraction of liquid B in the vapour is 0.673

4.16 Approximately 5 × 10^{-4} mol dm^3

4.17 The activity coefficients for A and B are 1.35 and 2.66 respectively

CHAPTER 5

5.1   (c) $-\dfrac{\Delta[H_2]}{\Delta t} = \dfrac{3}{2}\dfrac{\Delta[NH_3]}{\Delta t}$

5.2   (c) Rate $= -\dfrac{d[S_2O_8^{2-}]}{dt} = -\dfrac{1}{3}\dfrac{d[I^-]}{dt}$
      $= \dfrac{1}{2}\dfrac{d[SO_4^{2-}]}{dt} = \dfrac{d[I_3^-]}{dt}$

5.3   3 × 10^{-5} mol dm^{-3} s^{-1}

5.4   0.03 mol

5.5   (a) 2nd, 1st, and overall order = 3rd
      (b) $\dfrac{d[N_2]}{dt} = -\dfrac{1}{2}\dfrac{d[NO]}{dt}$
      (c) 25

5.6   (a) Rate = k[CH$_3$Cl][OH$^-$]
      (b) i. 1st,   ii. 1st
      (c) 2nd

5.7   9

5.8   Rate = k[ClO•][NO$_2$][N$_2$], k = 0.146 × 10^8 mol^{-2} dm^6 s^{-1}

5.9   3

5.10 B

5.11 (a) Rate$_2$:Rate$_1$ = 4:1
      (b) Rate$_2$:Rate$_1$ = 4:1
      (c) Rate$_2$:Rate$_1$ = 4:1

5.12 30 M^{-1} s^{-1}

5.13 2nd order and 0.00377 mol

5.14 8 mg

5.15 (a) 12 torr
      (b) 31 s

5.16 (c) k = 1.5 × 10^{-3} s

5.17 274 kJ mol^{-1}, 3.6 s^{-1}

5.18 $\dfrac{d[C]}{dt} = k_1[A]$

5.19 $\dfrac{d[O_2]}{dt} = k_2 \dfrac{k_f[N_2O_5]}{k_b + 2k_2}$

CHAPTER 6

6.1   (a) −4.6 × 10^{-20} N
      (b) −1.9 × 10^{-20} N

6.2   (a) 81.4 × 10^{-6}
      (b) 1.57 × 10^{-6}
      (c) i. 0.159 m,   ii. 0.0833,   iii. 576 × 10^{-6}

6.3   (a) 0.6 mol kg^{-1}
      (b) 1.9 mol dm^{-3}

6.4   (a) 0.495
      (b) 0.847

6.5   71.9 S cm^2 mol^{-1}

6.6   0.125, 0.0415, and 0.0128

6.8   (a) ii. +1.10 V,
         iii. −212 kJ mol^{-1},
         iv. 2 × 10^{37}
      (b) ii. +0.59 V,
         iii. −110 kJ mol^{-1},
         iv. 9 × 10^{19}
      (c) ii. +0.43 V,
         iii. −83 kJ mol^{-1},
         iv. 4 × 10^{14}

6.9   (a) +0.48 V
      (b) +0.29 V
      (c) −0.29 V
      (d) +0.73 V

6.10 (a) 48 × 10^{-3} g
      (b) 6300 s

Synoptic questions

CHAPTER 1

S1.1 92.4%
S1.2 43 kg
S1.3 a $C_3F_6OH_2$
b $N_2O = 0.625$ atm; $O_2 = 0.3125$ atm; desflurane = 0.0625 atm
S1.4 i 5.1×10^{18} kg
ii 1.8×10^{21} mol
S1.5 73.9%

CHAPTER 2

S2.1 c -1152 kJ mol^{-1}
S2.2 -2986 kJ mol^{-1}
S2.3 -60 kJ mol^{-1}
S2.4 a -3.716 kJ
b -3277 kJ mol^{-1}
S2.5 353 kJ mol^{-1} (from bond enthalpy values), 657 kJ mol^{-1} (from experimental values)

CHAPTER 3

S3.1 5.3 kJ mol^{-1}; 0.12 atm; 4.8×10^{-3} mol dm^{-3}
S3.2 92.5%
S3.3 1.1×10^{-27} mol dm^{-3}
S3.4 396 atm^2; 1.24 mol^2 dm^{-6}; 333×10^{-6} atm^{-2}
S3.5 116×10^{-6} mol dm^{-3}; 120.2×10^{-15} mol dm^{-3}

CHAPTER 4

S4.3 30 kJ mol^{-1}
S4.4 109 J K^{-1} mol^{-1}
S4.5 8.0×10^5 g mol^{-1}

CHAPTER 5

S5.1 ii. 0.086 day^{-1}
iii. 33 days
S5.2 21 kPa, 35 min
S5.3 902 kJ mol^{-1}
S5.4 i. 2.07×10^4 s
ii. 491 torr

CHAPTER 6

S6.1 0.459 and 5.15×10^{-4}
S6.2 113.7 S cm^2 mol^{-1}; 56.71 S cm^2 mol^{-1}/(mol dm^{-3})$^{1/2}$; 37.3 S cm^2 mol^{-1}
S6.3 $+0.76$ V
S6.4 0.118 g

Index

Numbers in brackets () indicate preliminary page numbers
Page numbers in bold type refer to sidenotes

A

acceleration
  definition/symbol (5)
  unit name (5)
acids
  dissociation constant 77
  equilibrium constant 76-7
activity coefficient
  real solutions 109
  synoptic questions 175
  worked example/solution 110
activity of electrolyte
  definition of 152
  'taking the geometric mean' **152**
  worked example/solution 153-4
allotropy
  elements having 89
  phase diagrams and 89
antilog function **78**
area
  definition/symbol (5)
  unit name (5)
Arrhenius equation
  Boltzmann distribution of energies in a gas 138-40
  worked examples/solutions 140-2
atomic mass unit
  average 6
  definition of **6**
  relative 5
Avogadro constant
  definition 5
  large number 2
  moles of atoms 5
  symbol/value (4), 5

B

bases
  equilibrium constant and 76-7
  boiling point of a substance **92**
Boltzmann constant
  distribution of energies in a gas 138-9
  symbol/value of (4)
bond enthalpies
  bond breaking/making **41**
  bond dissociation enthalpy 40
  mean bond enthalpy 40
  worked examples/solutions 40-2

Born-Haber cycle
  definition of 37
  lattice enthalpy 37
  worked example/solutions 38-40
Boyle's Law 19, 83

C

calorimetry 42-6
Celsius/Kelvin temperature scale 42-4
Charles's Law 19, 83
chemical equilibrium
  acids 76-7
  bases 77
  compositions, calculating 68, 69-74
  constants, calculating 68-74
  dynamic equilibrium 58
  forward reactions 63-4
  Gibbs energy minimum 58-9
  homogeneous 59
  heterogeneous 59
  Le Chatelier's principle 64-5
  mixed reagents 58
  relationship between different constants 62-3
  reverse reactions 63-4
  solubility 74-6
  standard Gibbs energy change 65-6
  temperature effect 66-8
  types of constants 60
  water 76-7
chemical reactions, half-life of
  first order reaction 132-3
  integrated rate equation 133
  second order reaction 133-7
  zero order reaction 132
  worked examples/solutions 134-7
chemical reactions, order of
  experimental measurements determining 137-8
  rate constant 118
  single-step reaction **118**
  worked example/solution 118-19
chemical reactions, rate of
  Arrhenius equation *see* Arrhenius equation
  average rate 114
  changes of rates 115-16
  definition of 114
  experimental measurements determining constants 137-8
  initial rate of reaction 115
  positive value 114
  rate dependent on temperature 138-40
  single-step reaction **118**

  working example/solution 114-15, 116, 117
chloralkali process 161
Clapeyron equation 94-5
Clausius-Clapeyron equation 95-8
colligative properties 83, 99
concentration
  definition/symbol (5)
  unit name (5)
conductivity
  electrolytes and 151
  Kohlrausch's law 157
  molar conductivity of electrolytes 157-60
  worked examples/solutions 158-60
Coulomb's law
  definition of 151
  worked example/solution 151-2

D

Dalton's Law of Partial Pressures 23
Daniell cell 160
Debye-Hückel limiting law
  activity coefficient for electrolytes using 156-7
  definition 156
  electrochemical process, use in 151
  working examples/solutions 156-7
density
  definition/symbol (5)
  gas *see below* density of gas
  unit name (5)
density of gas
  molar mass of gas 22
  worked example/solution 22
derived units
  table of (5)
dielectric constant
  solvents and **151**
  water and **152**
dimensionless quantity 59, 60, 151-4, 156, 160-1
Downs process 161
dynamic equilibrium 58

E

electric charge
  definition/symbol (5)
  unit name (5)
electrochemistry
  electrochemical cells *see* electrochemical cells
  electrolytes *see* electrolytes

INDEX

electrochemistry (*Cont.*)
  electrostatic interactions *see* electrostatic interactions
  electrochemical cells
    anodes 160
    cathodes 160
    electrolytic cell 161
    galvanic cells 160
    half reactions 160
    standard Gibbs energy 161
    standard potential, formulae to calculate 160-1
    standard potential of a cell 160-1, 164-5
    worked examples/solutions 163-9
electrolytes
  activity of 152-4
  Coulomb's law 151
  Debye-Hückel limiting law 156
  definition of 151
  degree of dissociation of 159-60
  dielectric constant **151**
  ionic strength 154-5
  mean activity coefficient 156-7
  molar conductivity of electrolytes 157-60
  strong 157
  variable conductivity of 151
electrolytic cell
  definition/examples of 161
electron mass
  symbol/value of (4)
electroplating 161
electrostatic interactions
  Coulomb's law 151
  dielectric constant **151**, **152**
  electrolytes *see* electrolytes
  worked example/solution 151-2
elementary charge
  symbol/value of (4)
empirical formula, determination of
  working example/solution 9-10
endothermic reaction
  bond breaking 41
  chemical equilibrium and 66
energy
  definition/symbol (5)
  unit name (5)
energy changes in thermodynamics
  matter/energy, definition of 25
  surroundings, definition of 25
  system *see* systems in thermodynamics
  universe, definition of 25
enthalpy changes
  bond enthalpies *see* bond enthalpies
  combustion and 45-6
  definition of 31, **31**
  fusion **94**, 96
  lattice enthalpy **37**
  sublimation 95
  vaporization **93**, 95, 97
  worked examples/solutions 31-3
entropy
  change in the surroundings, worked example/solution 50-1
  change with temperature, worked example/solution 48-9
  definition of 46
  determining change quantitatively 47-8
  direction of change, worked example/solution 47-8
  Gibbs free energy *see* Gibbs free energy
  molar **48**
  Second Law of Thermodynamics 47, 50, 52
  standard reaction, worked example/solution 49-50
  state function, as 46
equilibrium
  chemical *see* chemical equilibrium
  phase *see* phase equilbirium
exothermic reaction
  bond making 41
  chemical equilibrium and 66
extensive property
  thermodynamics and 26
  volume as 26

F

Faraday constant
  symbol/value of (4)
Faraday's laws of electrolysis 151, 161
First Law of Thermodynamics **27**, 33
first-order rate equations 124-5
force
  definition/symbol (5)
  unit name (5)
forward reactions
  chemical equilibrium and 63-4
  dynamic equilibrium and 58
frequency
  definition/symbol (5)
  unit name (5)
fugacity 85-6

G

galvanic cells
  definition/examples of 160
gas
  density of 22
  enthalpy changes *see* enthalpy changes
  expansion work, calculating 29-31
  ideal gas 20-1
  mixtures 23-4
gas expansion
  changing external pressure, work done against 29-31
  constant pressure, work done against 29
gas laws
  Avogadro's Law 19
  Boyle's Law 19
  Boyle's Law/Charles's Law combined 20
  Charles's law 19
  Dalton's Law of Partial Pressures 23
  density of gas 22
  ideal gas equation 20-1
  mixtures of gases 23
  variables in physical behaviour 19
  worked examples/solutions 20, 21, 22, 23,
gaseous phase
  Avogardo's law 83
  behaviour of 83
  Boyle's law 83
  Charles's law 83
  chemical potential of 85
  fugacity 85-6
  Gibbs energy of ideal gas, variations in 85
  ideal gas equation **83**
  ideal gas mixture 84
  phase equilibrium and 83-4
  real gases 85-6
  worked example/solution 87-8
Gibbs-Duhem equation 86
Gibbs energy minimum
  chemical equilibrium and 58-9
  standard change 65-6
Gibbs energy, standard
  half reactions and 160-1
  phase diagrams and 89
Gibbs free energy change
  equilibrium and 56-7, 65-6
  reaction isotherm 56
  Second Law of Thermodynamics 51-2
  spontaneous reactions 52
  worked examples/solutions 52-6
gravimetric analysis
  definition of 10
  worked examples/solutions 10-12

H

half reactions
  electrochemical series 160
  galvanic cells 160
  standard Gibbs energy 160, 161
  standard potentials 160
  worked examples/solutions 162
heat capacity
  calorimetry 44-5
  Celsius/Kelvin temperature scale 43
  definition of 42-3
  worked examples/solutions 43-5
Henry's Law 107
Hess's Law
  definition of 33
  enthalpy change in reaction, representation of 33
  worked examples/solutions 33-7

I

ideal dilute solution
  Raoult's law and 107
ideal gas
  calculating volume of 20-1
  equation/law 20-1

INDEX

ideal gas constant
　symbol/value of (4)
ideal gas equation 83
integrated rate equations
　definition of 124
　first-order rate equations 124-5
　second-order rate equations 126-32
　zero-order rate equations 124
intensive property
　concentration as 26
　density as 26
　pressure of gas 26
　thermodynamics and 26
internal energy
　expansion work done by gas against constant pressure 29
　expansion work done by gas against changing external pressure 29-31
　definition in thermodynamics 27
　First Law of Thermodynamics 27
　types of 27
　worked example/solution 28-9
ionic equation
　neutralization and 15
ionic strength of electrolyte
　definition of 154
　worked examples/solutions 154-5

J

joule, definition of 1

K

Kapustinskii constant
　symbol/value of (4)
kinetic energy of particle
　expression for 1
　type of internal energy, as 27
Kohlrausch's law 157

L

lattice enthalpy
　Born-Haber cycle 37-8
　direct measurement 38
　direction of measurement change **38**
　definition of **37**
　worked example/solution 38-40
Laws of Thermodynamics
　First Law **27**, 33
　Second Law 47, 50, 51
Le Chatelier's principle 64-5
Leclanché cell 160
Ligand substitution reactions
　stability constants 80-1
　working examples/solutions 81-2
limiting reagents
　definition of 18
liquid phase
　chemical potential of a 86
　Gibbs-Duhem equation 86

ideal solution 86
　gravitational forces 86
　real solutions 86
　vapour pressure 86

M

mean activity coefficient
　Debye-Hückel limiting law 156
　working examples/solutions 156-7
mixture of gases
　Dalton's Law of Partial Pressures 23
　worked examples/solution 23-4
molality 100, 101, **153**
molar conductivity
　concentration of electrolytes, effect of **157**, **158**
　limiting 157
　worked examples/solutions 158-60
molar mass
　definition 7
　determining amount of substance 7-8
　relative formula mass 7
　worked example/solution 7-8
molarity
　definition of 12
　equation to calculate 12
　worked examples/solutions 13
molecular formula
　determination of 9
　working example/solution 9-10
molecularity
　definition of 143
moles of atoms
　Avogadro's constant 5
　definition of 5
　relative atomic mass
　worked examples/solutions 5-6

N

natural logarithms **125**
normal logarithms **125**
Nernst equation 151, 161
neutron mass
　symbol/value of (4)
numbers, large and small
　Avogadro's number 2
　Planck constant 2
　worked examples/solution 2-3

O

osmotic pressure 100
one-component systems, phase equilibrium
　Clapeyron equation 94-5
　Clausius-Clapeyron equation 95-8
　enthalpy 92, 96
　entropy 92
　Gibbs energy 90-4
　molar Gibbs energy 90, 91

molar volume **91**
　phase behaviour see phase behaviour
　stable phase 90
　Trouton's rule 92, 93

P

path function
　definition of 26
percentage yield of a reaction
　definition of 18
　worked example/solution 18-19
permittivity of a vacuum 151
permittivity of free space
　symbol/value of (4)
periodic table
　relative atomic mass 5
　table of elements (3)
phase
　definition of 83
phase behaviour
　allotrophy 89
　Gibbs energy and 89, 90-1
　phase diagram 88-9
　phase transitions 88
　polymorphism 89
　stable phase, definition of 89
　worked example/solution 89-90
phase diagram
　allotrophy 89
　interpreting 88-9
　polymorphism 89
phase equilibrium
　gaseous see gaseous phase
　liquid see liquid phase
　one-component see one-component systems, phase equilibrium
　phase behaviour see phase behaviour
　phase, definition of 83
　phase diagram, interpreting 88-9
　phase transitions 88
　solid see solid phase
　two components see two components mixtures, phase equilibrium
phase rule 89
　phase transitions
　central thermodynamic equation for 88
physical constants (4)
Planck constant
　small number 2
　symbol/value of (4), 2
polymorphism
　examples of 89
　phase diagrams and 89
potential difference
　definition/symbol (5)
　unit name (5)
potential energy
　type of internal energy 27
　worked example/solution 1-2
power
　definition/symbol (5)
　unit name (5)

INDEX

pre-equilibrium kinetics 143
pressure
  definition/symbol (5)
  unit name (5)
proton mass
  symbol/value of (4)

R

radioisotope decay 132
Raoult's Law 98, 102, 107, 109
rate equations
  initial rates method for determining 120
  integrated *see* integrated rate equations
  worked equations/solutions 120-3
reaction isotherm 56
reaction kinetics
  half life of *see* chemical reactions, half life of
  order of *see* chemical reactions, order of
  rate of *see* chemical reactions, rate of
  rate equations *see* rate equations
  reaction mechanisms *see* reaction mechanisms
reaction mechanisms
  definition of 143
  elementary reactions defined 143
  molecularity 143
  rate dependence
  rate-determining step 143, 144
  mechanisms with reversible steps 143-9
  plausible mechanisms, worked example/solution 147-9
  pre-equilibrium kinetics 143
  pseudo-first-order 149
  short-lived intermediate kinetics 144-6
  steady-state approximation 143
redox titrations 15, 17
relative atomic mass 5, **7**
relative formula mass 7
relative numbers of atoms 9
reverse reactions
  chemical equilibrium and 63-4
  dynamic equilibrium and 58
reversible isothermal expansion of gas 29-31
rotational energy 27
Rydberg constant
  symbol/value of (4)

S

scientific notation, definition of **3**
Second Law of Thermodynamics 47, 50, 51
second-order rate equations
  rate equation of 126
  worked examples/solutions 126-32
short-lived intermediate kinetics
  reaction mechanisms with reversible steps 144
  worked examples/solutions 144-9
SI base units ((Systeme International d'Unites)

amount of substance,
  name/symbol(4), 1
  converting between units, worked examples/solutions 4-5
  electric current, name/symbol (4), 1
  length, name/symbol (4), 1
  luminous intensity, name/symbol (4)
  mass, name/symbol (4), 1
  temperature, name/symbol (4), 1
  time, name/symbol (4), 1
  worked examples/solutions 1-2
single-step reaction **118**
solid phase
  standard state of substance 87
solubility
  definition of 74
  equilibrium constant and 74
  worked examples/solutions 74-6
solutions
  concentrations of 12-13
  diluting solutions 14-15
  molarity 12
  stock solutions 14
  worked examples/solutions 13-14
speed of light in vacuum
  symbol/value of (4)
spontaneous reactions 52-4
stable phase, definition of 89
state function
  definition of 26
  entropy 46
  examples of 26
  internal energy as a 27
steady-state approximation 143
stock solutions 14
stoichiometric equation 11, 15, 26, 35, 49, 51, 55, **118**, 131, 132
synoptic questions 170-5
systems in thermodynamics
  closed 25
  definition of 25
  examples of 25-6
  extensive property 26
  intensive property 26
  isolated 25
  internal energy 27-9
  matter/energy transfers in 25
  open 25
  path function as property of 26
  state function as property of 26
  surrounding, definition of 25
  worked examples/solutions 25-6

T

'taking the geometric mean' **152**
theoretical yield of a reaction
  definition of 18
thermodynamics
  bond enthalpies *see* bond enthalpies
  Born-Haber cycle 37-40
  energy changes *see* energy changes in thermodynamics
  enthalpy changes *see* enthalpy changes

entropy *see* entropy
heat capacity *see* heat capacity
Hess's Law 33
internal energy *see* internal energy
laws of *see* Laws of Thermodynamics
state function 26
systems *see* systems in thermodynamics
titrations 15, 17
Trouton's rule 92, 93
two-component mixtures, phase equilibrium
  azeotropes 113
  collagative properties 99
  distillation of binary liquid mixture 111-13
  Henry's law 107
  ideal binary liquid mixtures 102-7
  ideal dilute solutions 107
  non-ideal binary liquid mixtures 109-10
  non-volatile solute/volatile solvent 98-102
  partial miscibility 113
  Raoult's law 98, 102, 107, 109, 113

U

units
  converting in chemical calculations 4
  multiples of (5)
  prefixes of (5), 2
  worked examples/solutions 4-5
units of pressure
  relationships between 21

V

vapour 83
vapour pressure 83, 86, 98, 105
velocity
  definition/symbol (5)
  unit name (5)
vibrational energy 27
volume
  definition/symbol (5)
  unit name (5)
volumetric analysis
  acids, mono/di/tribasic 15
  ionic equation for neutralization 15
  titrations 15
  worked examples/solutions 15-17

W

water
  amphiprotic nature of 77
  dielectric constant **151**
  equilibrium constant 77
  solvent for electrolytes, as **151**

Z

zero-order rate equations 124